Antennas and Wave Propagation

Antennas and Wave Propagation

Chris Harvey

CLANRYE
INTERNATIONAL
www.clanryeinternational.com

Clanrye International,
750 Third Avenue, 9th Floor,
New York, NY 10017, USA

ISBN: 978-1-64726-145-0

Cataloging-in-Publication Data

Antennas and wave propagation / Chris Harvey.
p. cm.
Includes bibliographical references and index.
ISBN: 978-1-64726-145-0
1. Antennas (Electronics). 2. Radio wave propagation. 3. Electromagnetic waves.
4. Electronic apparatus and appliances. 5. Wave-motion, Theory of. I. Harvey, Chris.
TK7871.6 .A58 2022
621.382 4--dc23

For information on all Clanrye International publications
visit our website at www.clanryeinternational.com

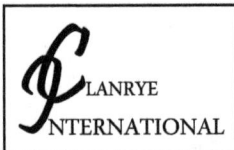

CLANRYE
INTERNATIONAL

Table of Contents

Preface

It is with great pleasure that I present this book. It has been carefully written after numerous discussions with my peers and other practitioners of the field. I would like to take this opportunity to thank my family and friends who have been extremely supporting at every step in my life.

Antenna is an array of conductors. It is the interface between radio waves which propagate through electric and space currents in metal conductors. They are required by transmitters and radio receivers to combine its electrical connection to electromagnetic field. Radio waves are electromagnetic waves. They carry signals at the speed of light through air without any transmission loss. They can be classified by operating principles or applications. Antennas are classified as omnidirectional or directional. Other types include whip antenna, dipole antenna, etc. Antennas and propagation act as keys for any radio system. Wave propagation is the study of the ways in which waves travel. The study of radio wave's behavior while traveling from one point to another is known as radio propagation. Most of the topics introduced in this book cover new techniques and the applications of antennas and wave propagation. It aims to shed light on some of the unexplored aspects of this field. It will serve as a valuable source of reference for those interested in antennas and wave propagation.

The chapters below are organized to facilitate a comprehensive understanding of the subject:

Chapter – Antenna: An Introduction

Antenna is defined as the transducer that converts radio frequencies into alternating current and optimizes the matching system between the sources of electromagnetic energy and space. Different types of antennas are dipole antenna, loop antenna, aperture antenna, reflector antenna, etc. This is an introductory chapter which will briefly introduce antenna and its types.

Chapter – Antenna: Parameters and Concepts

Some of the significant concepts and parameters of antenna are beam width, antenna aperture, antenna factor, antenna gain, antenna impedance, radiation pattern, antenna polarization, impedance matching, etc. This chapter has been carefully written to provide an easy understanding of these concepts and parameters of antenna.

Chapter – Wire Antennas

Wire antenna is a form of antenna that consists of a long wire whose length does not depend on the wavelength of the radio waves. A few aspects that are studied in relation to wire antennas are transmission lines, short dipole, monopole, etc. The topics elaborated in this chapter will help in gaining a better perspective about wire antennas.

Chapter – Aperture Antennas

The antenna with an aperture at the edge of a transmission line which radiates energy is known as aperture antenna. A few of its types are slot antenna, parabolic antenna, horn antenna, etc. This chapter closely examines these types of aperture antenna to provide an extensive understanding of the subject.

Chapter – Antenna Arrays

A collective network of multiple antennas which works together as a single unit for propagation of radio waves is termed as an antenna array. A few of its types include collinear array, parasitic array, phased array, curtain array, end-fire array, etc. All these types of antenna arrays have been carefully analyzed in this chapter.

Chapter – Wave Propagation

The different ways in which waves travel can be referred to as wave propagation. Sky wave propagation, space wave propagation, ionospheric propagation, ground wave propagation, radio propagation, etc. are some of the ways in which waves propagate. This chapter has been carefully written to provide an easy understanding of wave propagation.

Chris Harvey

1

Antenna: An Introduction

Antenna is defined as the transducer that converts radio frequencies into alternating current and optimizes the matching system between the sources of electromagnetic energy and space. Different types of antennas are dipole antenna, loop antenna, aperture antenna, reflector antenna, etc. This is an introductory chapter which will briefly introduce antenna and its types.

An antenna is used to radiate electromagnetic energy efficiently and in desired directions. Antennas act as matching systems between sources of electromagnetic energy and space. The goal in using antennas is to optimize this matching. Here is a list of some of the properties of antennas:

1. Field intensity for various directions (antenna pattern).

2. Total power radiated when the antenna is excited by a current or voltage of known intensity.

3. Radiation efficiency which is the ratio of power radiated to the total power.

4. The input impedance of the antenna for maximum power transfer (matching).

5. The bandwidth of the antenna or range of frequencies over which the above properties are nearly constant.

All antennas may be used to receive or radiate energy.

Different Types of Antennas

1. Dipole Antennas: The dipole is one of the most common antennas. It consists of a straight conductor excited by a voltage from a transmission line or a waveguide. Dipoles are easy to make.

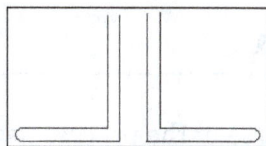
Dipole Antenna

2. Loop Antennas: A loop of wire, with many turns, is used to radiate or receive electromagnetic energy.

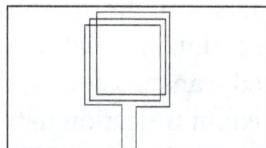
Loop Antenna

3-Aperture Antennas: A horn as shown in the figure below is an example of an aperture antenna. These types of antennas are used in aircraft and spacecraft.

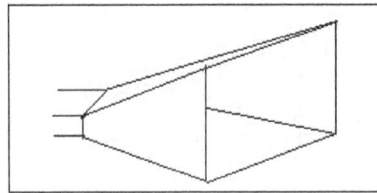

Horn antenna.

4-Reflector Antennas: The parabolic reflector is a good example of reflectors at microwave frequencies. In the past, parabolic reflectors were used mainly in space applications but today they are very popular and are used by almost everyone who wishes to receive the large number of television channels transmitted all over the globe.

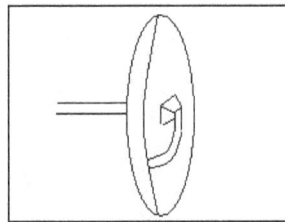

Parabolic reflector with feed at focus.

5-Array Antennas: A grouping of similar or different antennas form an array antenna. The control of phase shift from element to element is used to scan electronically the direction of radiation.

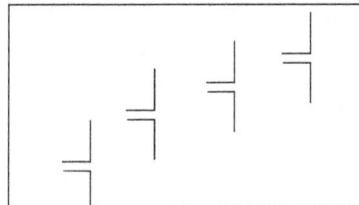

Array Antennas.

RECEIVING ANTENNA

A receiving antenna performs the reverse of the process performed by the transmission antenna. It receives radiofrequency radiation and converts it into electric currents in an electric circuit connected to the antenna. TV and radio broadcasting stations, use transmission antennas to transmit specific types of signals that propagate through the air. These signals are detected by receiving antennas which convert them into signals, and are received by the appropriate device (e.g., TV, radio, mobile phone). Cellular communication installations (e.g. base stations, repeaters) and mobile phones are equipped with designated transmission and receiving antennas that emit radiofrequency radiation and serve the cellular communication networks, in accordance with communication network technologies.

Equivalent Circuit Model of a Radio Link

To study the behaviour of a receiving antenna, we will consider a link consisting of two antennas: one transmitting and one receiving. We wish to understand the behaviour of such a system. Since an antenna is a one-port device, the analysis is facilitated by considering the system of antennas to be a 2-port "black box" with unknown internal characteristics.

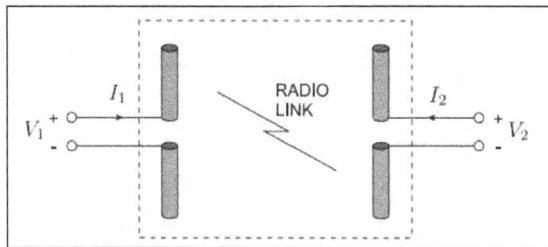

Recall from circuit theory than an unknown "black box" with two ports is fully characterized if we know the terminal voltage (V_1, V_2) and currents (I_1, I_2) of the device. These quantities are related through.

$$V_1 = Z_{11}I_1 + Z_{12}I_2$$
$$V_2 = Z_{21}I_1 + Z_{22}I_2,$$

or simply,

$$[V] = [Z][I],$$

where [Z] is the impedance matrix of the two-port network. Z_{11} and Z_{22} are called self-impedances of the system while Z_{12} and Z_{21} are called mutual impedances of the network. Each of the impedances can be found through open-circuiting ports of the network, such that,

$$Z_{mn} = \frac{V_m}{I_n}\bigg|_{I_k=0 \text{ for } k \neq n}$$

for an arbitrary network.

Note that the input impedance seen looking into one port is a function of the loading on the second antenna. For example, the input impedance seen looking into port 1 is,

$$Z_1 = \frac{V_1}{I_1} = \frac{Z_{11}I_1 + Z_{12}I_2}{I_1} = Z_{11} + \frac{Z_{12}I_2}{I_1}$$

Hence, the current flowing on antenna 2 does influence the input impedance seen looking into antenna 1. However, if antenna 2 is placed very far away, the influence of the second antenna should be negligible ($Z_{12} \approx 0$), yielding $Z_1 \approx Z_{11}$, the self-impedance of antenna 1.

We are most concerned with the mutual impedance terms of our circuit model, since they describe the coupling between antenna 1 and antenna 2 (and vice versa, if antenna 2 was connected to the source). But first, we must consider a very important theorem that describes the behaviour of the system we have described when the antennas are immersed in a homogeneous, linear, passive, and isotropic medium (like free space).

Reciprocity

One of the most important electromagnetic concepts is that of reciprocity, which is the behaviour of an electromagnetic system in a simple medium. Consider the situation shown below, where we have a volume containing to sources, J_1 and J_2, which each produce fields E_1, H_1 and E_2, H_2, respectively, as shown in figure.

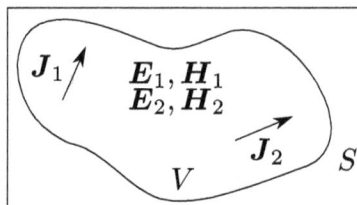

Volume containing two electric sources

Mathematically, reciprocity was developed by Lorentz but a useful form of that theorem for our purposes states that,

$$\iiint_V J_1 \cdot E_2 \, dv' = \iiint_V J_2 \cdot E_1 \, dv'.$$

Now consider the two-antenna (two-source) problem we considered originally, but the antennas located a large distance apart such that the assumptions above hold. Let's say the two antennas are ideal dipole antennas, driven by ideal current generators (with infinite source impedances).

Thinking of the structure of the antenna, E_2, the field produced at antenna 1 by antenna 2 (source J_2) only has a nonzero projection with J_1 across the antenna terminals, since J_1 is only nonzero along the conducting part of the antenna, and E_2 is only nonzero across the gap because of PEC boundary conditions. If we assume that the current J_1 can be represented as a linear current (existing only over a contour instead of a volume) and furthermore the current is uniform across the gap,

$$\iiint_V E_2 \cdot J_1 = \int_C E_2 I_1 dl = I_1 \int_C E_2 dl = -V_1^{OC} I_1,$$

where V_1^{OC} is the open-circuited voltage produced at antenna 1 as a result of the incident field produced by antenna 2 (E_2). Similarly, at antenna 2,

$$\iiint_V E_1 \cdot J_2 = -V_2^{OC} I_2$$

According to the Reciprocity Theorem, last two equations must be equal:

$$V_1^{OC} I_1 = V_2^{OC} I_2$$

Or

$$\frac{V_1^{OC}}{I_2} = \frac{V_2^{OC}}{I_1}$$

From our equivalent circuit model of a two-antenna link, we know that,

$$Z_{21} = \left.\frac{V_2}{I_1}\right|_{I_2=0} = \frac{V_2^{OC}}{I_1}$$

and

$$Z_{12} = \left.\frac{V_1}{I_2}\right|_{I_1=0} = \frac{V_1^{OC}}{I_2},$$

therefore,

$$Z_{12} = Z_{21}$$

This is the fundamental definition of a reciprocal two-port circuit, because we see that if we excite port 1 with a current source of amplitude I, the open circuit-voltage at port 2 is Z_{21}I while if we flip the current source to port 2, the open-circuit voltage at port 1 is Z_{12}I = Z_{21}I which is the same result as with the current source at port 1. That is, if we drive port 1 with an ideal current source having amplitude I, the open circuit voltage at port 2 is:

$$V_2^{OC} = Z_{21}I.$$

If we flip the current source to port 2, then the open circuit voltage at port 1 is:

$$V_1^{OC} = Z_{12}I.$$

Hence, in a reciprocal system, both cases should produce the same open-circuit voltage, regardless of which antenna is transmitting and which is receiving.

The consequence of this on antennas is found as follows. Consider an experiment where we measure the transmit pattern of antenna (1), using a second receiving antenna (2) moving about a circle of fixed radius about the transmitting antenna while remaining co-polarized with the transmission, as shown in figure.

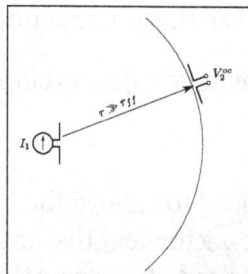

Antenna 1 transmits, antenna 2 receives.

If we measure the open-circuit voltage at antenna 2, we know it is equal to,

$$V_2^{OC}(\theta) = Z_{21}(\theta)I.$$

Now consider a second experiment where antenna (2) is used as the transmitter and antenna (1) as the receiver, as shown in figure. Antenna (2) moves in an identical manner as the first experiment, while this time we measure the open-circuit voltage at the terminals of antenna 1, which is equal to,

$$V_1^{OC}(\theta) = Z_{12}(\theta)I = Z_{21}(\theta)I.$$

Hence we see the "open-circuit voltage" pattern measured by both experiments is identical. Therefore, we conclude that the transmit and receive patterns of an antenna are the same.

Vector Effective Length

A receiving antenna is used to collect electromagnetic waves and extract power from them. The concept of the effective length of an antenna is used to determine the voltage induced on the open-circuited terminals of the antenna when a wave impinges on it. This effective length is a vector quantity and is defined as,

$$\ell_{\textit{eff}}(\theta,\phi) = \ell_\theta(\theta,\phi)\hat{\theta} + \ell_\phi(\theta,\phi)\hat{\phi}.$$

It is related to the far-zone electric field radiated by the antenna through,

$$E_{rad} = -\frac{j\omega\mu I}{4\pi}\frac{e^{-jkr}}{r}\ell_{\textit{eff}}(\theta,\phi).$$

In general, for a transmitting antenna, vector effective length can be defined as follows:

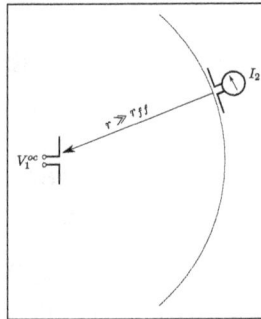

Antenna 2 transmits, antenna 1 receives

The vector effective length of an antenna is the length and orientation of a uniform current required to produce the same electric field as the antenna under consideration.

If the antenna is receiving, the open-circuit voltage developed across the antenna terminals is,

$$V^{OC} = E^i . \ell_{\textit{eff}}$$

We note that the complex conjugate is used to change the reference direction for the receiver (this only applies to antennas with complex vector lengths: those with elliptical or circular polarization). When $\ell_{\textit{eff}}$ and E i are linearly polarized, $\ell_{\textit{eff}}$ can be thought of as the vector length of a linear antenna that the open circuit voltage is being induced into by E^i. The radiated electric field of an ideal dipole oriented along the z-axis is, as we know,

$$E = \frac{j\omega\mu I\Delta z}{4\pi}\frac{e^{-jkr}}{r}\sin\theta\,\hat{\theta}.$$

Comparing Equations $E = \frac{j\omega\mu I\Delta z}{4\pi}\frac{e^{-jkr}}{r}\sin\theta\,\hat{\theta}$ and $E_{rad} = -\frac{j\omega\mu I}{4\pi}\frac{e^{-jkr}}{r}\ell_{eff}(\theta,\varphi),$

$$\ell_{eff} = -\Delta z\sin\theta\,\hat{\theta}$$

produces $E = \frac{j\omega\mu I\Delta z}{4\pi}\frac{e^{-jkr}}{r}\sin\theta\,\hat{\theta}.$ Hence, Equation $\ell_{eff} = -\Delta z\sin\theta\,\hat{\theta}$ represents the vector effective length of an ideal dipole oriented along the z-axis.

Example: Half-wave dipole along the z-axis.

$$E = j\omega\mu\frac{2I}{4\pi k}\frac{e^{-jkr}}{r}\frac{\cos\left(\frac{\pi}{2}\cos\theta\right)}{\sin\theta}\hat{\theta}$$

$$= \frac{j\omega\mu}{4\pi}\frac{e^{-jkr}}{r}\left(\frac{2}{k}\frac{\cos\left(\frac{\pi}{2}\cos\theta\right)}{\sin\theta}\hat{\theta}\right)$$

$$\Rightarrow \ell_{eff} = -\frac{\lambda}{\pi}\frac{\cos\left(\frac{\pi}{2}\cos\theta\right)}{\sin\theta}\hat{\theta}.$$

Comparing the maximum effective length of a λ/2 dipole to that of an ideal dipole (i.e. for θ = 90°), we find that,

$$\frac{\left|\ell_{eff,\lambda/2}\right|}{\ell_{eff,ideal}} = \frac{\lambda}{\pi} = \frac{2}{\pi}\frac{\lambda}{2},$$

indicating that the effective length of a half-wave dipole corresponds to only (2/π = 63.7%) of its actual length (in contrast to an ideal dipole having $\ell_{eff} = \Delta\ell$)

Example: What is the induced voltage in an ideal dipole along the z-axis if the incident electric field is that of a plane wave travelling such that its k-vector makes an angle θ with the z-axis, and E-field is contained in the yz-plane and points towards the +z axis?

$$V^{OC} = E^i \cdot \ell_{eff}$$

$$= \left(-E_\theta\hat{\theta} + E_\phi\hat{\phi}\right)\cdot\left(-\Delta z\sin\theta\right)\hat{\theta}$$

$$= E_\theta\Delta z\sin\theta$$

We note the sign of V^{oc} is correct with respect to terminal conventions.

Example: What is the open circuit voltage magnitude developed at an ideal dipole that is perfectly aligned with the incident field, and aligned for maximum output?

$$V_{max}^{oc} = \max\left(\left|E^i \cdot \ell_{eff}\right|\right)$$

$$= \max E^i \ell_{eff}$$

$$= E^i \Delta\ell \max\left(\sin\theta\right)$$

$$= E^i \Delta\ell$$

Hence, for a perfectly aligned ideal dipole the terminal voltage is simply the product of the incident electric field along the physical length of the antenna. This is only true for ideal dipoles; hence, we could define vector effective length as the length of an ideal dipole that relates the open circuit voltage and incident field through this simple relationship.

We will use this result in the derivation of the relationship between an antenna's effective area and its gain.

To conclude, the concept of the vector effective length of an antenna is useful for two things:

1. It allows us to relate an incident electric field on any antenna to the open-circuit voltage developed at its terminals; and

2. It is a useful tool in determining the effect of polarization mismatch between the incident field and the antenna, which will be discussed later in the course.

2

Antenna: Parameters and Concepts

Some of the significant concepts and parameters of antenna are beam width, antenna aperture, antenna factor, antenna gain, antenna impedance, radiation pattern, antenna polarization, impedance matching, etc. This chapter has been carefully written to provide an easy understanding of these concepts and parameters of antenna.

BEAM WIDTH

Beam width is the aperture angle from where most of the power is radiated. The two main considerations of this beam width are Half Power Beam Width (HPBW) and First Null Beam Width (FNBW).

Half-power Beam Width

According to the standard definition, "The angular separation, in which the magnitude of the radiation pattern decreases by 50% (or -3dB) from the peak of the main beam, is the Half Power Beam Width."

In other words, Beam width is the area where most of the power is radiated, which is the peak power. Half power beam width is the angle in which relative power is more than 50% of the peak power, in the effective radiated field of the antenna.

Indication of HPBW

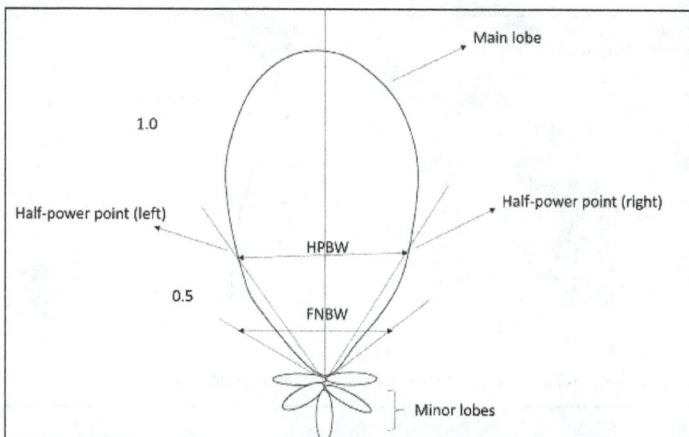

The figure shows half-power points on the major lobe and HPBW.

When a line is drawn between radiation pattern's origin and the half power points on the major lobe, on both the sides, the angle between those two vectors is termed as HPBW, half power beam width.

Mathematical Expression

The mathematical expression for half power beam width is –

Half power beam with = 70λ/D

Where

- λ is wavelength (λ = 0.3/frequency).
- D is Diameter.

Units

The unit of HPBW is radians or degrees.

First Null Beam Width

According to the standard definition, "The angular span between the first pattern nulls adjacent to the main lobe, is called as the First Null Beam Width."

Simply, FNBW is the angular separation, quoted away from the main beam, which is drawn between the null points of radiation pattern, on its major lobe.

Indication of FNBW

Draw tangents on both sides starting from the origin of the radiation pattern, tangential to the main beam. The angle between those two tangents is known as First Null Beam Width (FNBW).

This can be better understood with the help of the following diagram.

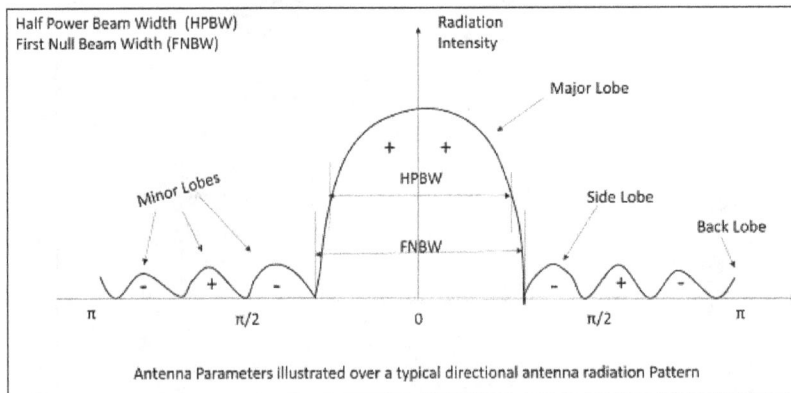

Antenna Parameters illustrated over a typical directional antenna radiation Pattern

The above image shows the half power beam width and first null beam width, marked in a radiation pattern along with minor and major lobes.

Mathematical Expression

The mathematical expression of First Null Beam Width is:

FNBW = 2 HPBW

FNBW 2 (70 λ/D) =140 λ/D

Where

- λ is wavelength (λ = 0.3/frequency).

- D is Diameter.

Units

The unit of FNBW is radians or degrees.

ANTENNA BANDWIDTH

Bandwidth is a fundamental antenna parameter. Bandwidth describes the range of frequencies over which the antenna can properly radiate or receive energy. Often, the desired bandwidth is one of the determining parameters used to decide upon an antenna.

The frequency range or *bandwidth* over which an antenna functions well can be very wide (as in a log-periodic antenna) or narrow (as in a small loop antenna); outside this range the antenna impedance becomes a poor match to the transmission line and transmitter (or receiver). Use of the antenna well away from its design frequency affects its radiation pattern, reducing its directive gain.

Generally an antenna will not have a feed-point impedance that matches that of a transmission line; a matching network between antenna terminals and the transmission line will improve power transfer to the antenna. The matching network may also limit the usable bandwidth of the antenna system. It may be desirable to use tubular elements, instead of thin wires, to make an antenna; these will allow a greater bandwidth. Or, several thin wires can be grouped in a *cage* to simulate a thicker element. This widens the bandwidth of the resonance.

Amateur radio antennas that operate at several frequency bands which are widely separated from each other may connect elements resonant at those different frequencies in parallel. Most of the transmitter's power will flow into the resonant element while the others present a high impedance. Another solution uses *traps*, parallel resonant circuits which are strategically placed in breaks along each antenna element. When used at one particular frequency band the trap presents a very high impedance (parallel resonance) effectively truncating the element at that length, making it a proper resonant antenna. At a lower frequency the trap allows the full length of the element to be employed, albeit with a shifted resonant frequency due to the inclusion of the trap's net reactance at that lower frequency.

The bandwidth characteristics of a resonant antenna element can be characterized according to its *Q* where the resistance involved is the radiation resistance, which represents the emission of energy from the resonant antenna to free space.

The Q of a narrow band antenna can be as high as 15. On the other hand, the reactance at the same off-resonant frequency of one using thick elements is much less, consequently resulting in a Q as low as 5. These two antennas may perform equivalently at the resonant frequency, but the second antenna will perform over a bandwidth 3 times as wide as the antenna consisting of a thin conductor.

Antennas for use over much broader frequency ranges are achieved using further techniques. Adjustment of a matching network can, in principle, allow for any antenna to be matched at any frequency. Thus the small loop antenna built into most AM broadcast (medium wave) receivers has a very narrow bandwidth, but is tuned using a parallel capacitance which is adjusted according to the receiver tuning. On the other hand, log-periodic antennas are *not* resonant at any frequency but can be built to attain similar characteristics (including feedpoint impedance) over any frequency range. These are therefore commonly used (in the form of directional log-periodic dipole arrays) as television antennas.

Antenna Systems

In the field of antennas, two different methods of expressing relative bandwidth are used for narrowband and wideband antennas. For either, a set of criteria is established to define the extents of the bandwidth, such as input impedance, pattern, or polarization.

Percent bandwidth, usually used for narrowband antennas, is used defined as:

$$\%B = 100 \times \frac{f_H - f_L}{f_c} = 200 \times \frac{f_H - f_L}{f_H + f_L}.$$

The theoretical limit to percent bandwidth is 200%, which occurs for $f_L = 0$.

Fractional bandwidth or ratio bandwidth, usually used for wideband antennas, is defined as $B = f_H / f_L$ and is typically presented in the form of $B:1$. Fractional bandwidth is used for wideband antennas because of the compression of the percent bandwidth that occurs mathematically with percent bandwidths above 100%, which corresponds to a fractional bandwidth of 3:1.

If $\%B = 200 \times \dfrac{f_H - f_L}{f_H + f_L} = p$

then $B = \dfrac{200 + p}{200 - p}$

ANTENNA APERTURE

In electromagnetics and antenna theory, antenna aperture, effective area, or receiving cross section, is a measure of how effective an antenna is at receiving the power of electromagnetic radiation (such as radio waves). The aperture is defined as the area, oriented perpendicular to the direction of an incoming electromagnetic wave, which would intercept the same amount of power from that wave as is produced by the antenna receiving it. At any point x, a beam of electromagnetic radiation has an *irradiance* or *power flux density* $S(x)$ which is the amount of power passing through a unit area of one square meter. If an antenna delivers P_o watts to the load connected to its output terminals (e.g. the receiver) when irradiated by a uniform field of power density S watts per square meter, the antenna's aperture A_e in square meters is given by:

$$A_e = \frac{P_o}{S}.$$

So the power received by an antenna (in watts) is equal to the power density of the electromagnetic energy (in watts per square meter), multiplied by its aperture (in square meters). The larger an antenna's aperture, the more power it can collect from a given electromagnetic field. To actually obtain the predicted power available P_o, the polarization of the incoming waves must match the polarization of the antenna, and the load (receiver) must be impedance matched to the antenna's feedpoint impedance.

Although this concept is based on an antenna receiving an electromagnetic wave, knowing A_e directly supplies the (power) gain of that antenna. Due to reciprocity, an antenna's gain in receiving and transmitting are identical. Therefore, A_e can be used to compute the performance of a transmitting antenna also. Note that A_e is a function of the direction of the electromagnetic wave relative to the orientation of the antenna, since the gain of an antenna varies according to its radiation pattern. When no direction is specified, is understood to refer to its maximum value, with the antenna oriented so its main lobe, the axis of maximum sensitivity, is directed toward the source.

Aperture Efficiency

In general, the aperture of an antenna is not directly related to its physical size. However some types of antennas, for example parabolic dishes and horn antennas, have a physical aperture (opening) which collects the radio waves. In these *aperture antennas*, the effective aperture A_e is always less than the area of the antenna's physical aperture A_{phys}, otherwise the antenna could produce more power from its terminals than the radio power entering its aperture, violating conservation of energy. An antenna's *aperture efficiency*, e_a is defined as the ratio of these two areas:

$$e_a = \frac{A_e}{A_{\text{phys}}}.$$

The aperture efficiency is a dimensionless parameter between 0 and 1.0 that measures how close the antenna comes to using all the radio wave power entering its physical aperture. If the antenna were perfectly efficient, all the radio power falling within its physical aperture would be converted to electrical power delivered to the load attached to its output terminals, so these two areas would be equal $A_e = A_{\text{phys}}$ and the aperture efficiency would be 1.0. But all antennas have losses, such as power dissipated as heat in the resistance of its elements, nonuniform illumination by its feed, and radio waves scattered by structural supports and diffraction at the aperture edge, which reduce the power output. Aperture efficiencies of typical antennas vary from 0.35 to 0.70 but can range up to 0.90.

Aperture and Gain

The directivity of an antenna, its ability to direct radio waves in one direction or receive from a single direction, is measured by a parameter called its isotropic gain G, which is the ratio of the power P_o received by the antenna to the power P_{iso} that would be received by a hypothetical isotropic antenna, which receives power equally well from all directions. It can be seen that gain is also equal to the ratio of the apertures of these antennas,

$$G = \frac{P_o}{P_{iso}} = \frac{A_e}{A_{iso}}.$$

the aperture of a lossless isotropic antenna, which by definition has unity gain, is

$$A_{iso} = \frac{\lambda^2}{4\pi}$$

where, λ is the wavelength of the radio waves. So

$$G = \frac{A_e}{A_{iso}} = \frac{4\pi A_e}{\lambda^2}$$

and for an antenna with a physical aperture of area A_{phys}

$$G = \frac{4\pi A_{phys} e_a}{\lambda^2}.$$

So antennas with large effective apertures are high gain antennas, which have small angular beam widths. As receiving antennas, they are most sensitive to radio waves coming from one direction, and are much less sensitive to waves coming from other directions. As transmitting antennas, most of their power is radiated in a narrow beam in one direction, and little in other directions. Although these terms can be used as a function of direction, when no direction is specified, the gain and aperture are understood to refer to the antenna's axis of maximum gain, or boresight.

Friis Transmission Formula

The fraction of the power delivered to a transmitting antenna that is received by a receiving antenna is proportional to the product of the apertures of both the antennas and inversely proportional to the distance between the antennas and wavelength. This is given by a form of the Friis transmission formula:

$$\frac{P_r}{P_t} = \frac{A_r A_t}{d^2 \lambda^2}$$

where:

- P_t is the power fed into the transmitting antenna input terminals;
- P_r is the power available at receiving antenna output terminals;
- A_r is the effective area of the receiving antenna;

- A_t is the effective area of the transmitting antenna;

- d is the distance between antennas. The formula is only valid for d large enough to ensure a plane wave front at the receive antenna, sufficiently approximated by $d \geq 2a^2 / \lambda$ where a is the largest linear dimension of either of the antennas.

- λ is the wavelength of the radio frequency;

The variables A_r, A_t, d, and λ must be expressed in the same units of length, such as meters, and variables P_t and P_r must be in the same units of power, such as watts.

Thin Element Antennas

In the case of thin element antennas such as monopoles and dipoles, there is no simple relationship between physical area and effective area. However, the effective areas can be calculated from their power gain figures:

Wire antenna	Power gain	Effective area
Short dipole (Hertzian dipole)	1.5	$0.1194\ \lambda^2$
Half-wave dipole	1.64	$0.1305\ \lambda^2$
Quarter-wave monopole	3.28	$0.2610\ \lambda^2$

This assumes that the monopole antenna is mounted above an infinite ground plane and that the antennas are lossless. When resistive losses are taken into account, particularly with small antennas, the antenna gain might be substantially less than the directivity, and the effective area is less by the same factor.

Effective Length

For antennas which are not defined by a physical area, such as monopoles and dipoles consisting of thin rod conductors, the aperture bears no obvious relation to the size or area of the antenna. An alternate measure of antenna gain that has a greater relationship to the physical structure of such antennas is *effective length* l_{eff} measured in metres, which is defined for a receiving antenna as:

$$l_{eff} = V_0 / E_s$$

where,

- V_0 is the open circuit voltage appearing across the antenna's terminals.

- E_s is the electric field strength of the radio signal, in volts per metre, at the antenna.

The longer the effective length the more voltage and therefore the more power the antenna will receive. Note, however, that an antenna's gain or A_{eff} increases according to the *square* of l_{eff}, and that this proportionality also involves the antenna's radiation resistance. Therefore, this measure is of more theoretical than practical value and is not, by itself, a useful figure of merit relating to an antenna's directivity.

Derivation of Aperture of an Isotropic Antenna

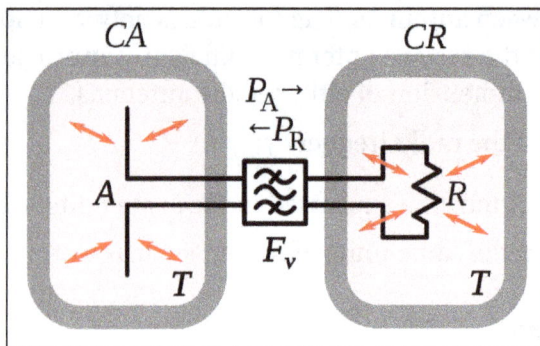

Diagram of antenna A and resistor R in thermal cavities, connected by filter F_ν.
If both cavities are at the same temperature T, $P_A = P_R$

The aperture of an isotropic antenna, the basis of the definition of gain above, can be derived by a thermodynamic argument. Suppose an ideal (lossless) isotropic antenna A located within a thermal cavity CA, is connected via a lossless transmission line through a band-pass filter F_ν to a matched resistor R in another thermal cavity CR (the characteristic impedance of the antenna, line and filter are all matched). Both cavities are at the same temperature T. The filter F_ν only allows through a narrow band of frequencies from ν to $\nu + \Delta\nu$. Both cavities are filled with black-body radiation in equilibrium with the antenna and resistor. Some of this radiation is received by the antenna. The amount of this power P_A within the band of frequencies $\Delta\nu$ passes through the transmission line and filter F_ν and is dissipated as heat in the resistor. The rest is reflected by the filter back to the antenna and is reradiated into the cavity. The resistor also produces Johnson–Nyquist noise current due to the random motion of its molecules at the temperature T. The amount of this power P_R within the frequency band ν to $\nu + \Delta\nu$ passes through the filter and is radiated by the antenna. Since the system is at a common temperature it is in thermodynamic equilibrium; there can be no net transfer of power between the cavities, otherwise one cavity would heat up and the other would cool down in violation of the second law of thermodynamics. Therefore, the power flows in both directions must be equal:

$$P_A = P_R.$$

The radio noise in the cavity is unpolarized, containing an equal mixture of polarization states. However any antenna with a single output is polarized, and can only receive one of two orthogonal polarization states. For example, a linearly polarized antenna cannot receive components of radio waves with electric field perpendicular to the antenna's linear elements; similarly a right circularly polarized antenna cannot receive left circularly polarized waves. Therefore, the antenna only receives the component of power density S in the cavity matched to its polarization, which is half of the total power density:

$$S_{matched} = \frac{1}{2}S.$$

Suppose B_ν is the spectral radiance per hertz in the cavity; the power of black body radiation per unit area (meter²) per unit solid angle (steradian) per unit frequency (hertz) at frequency ν and temperature T in the cavity. If $A_e(\theta,\phi)$ is the antenna's aperture, the amount of power in the frequency range Δv the antenna receives from an increment of solid angle $d\Omega = d\theta d\phi$ in the direction θ,ϕ is:

$$dP_A(\theta,\phi) = A_e(\theta,\phi)S_{matched}\Delta v d\Omega = \frac{1}{2}A_e(\theta,\phi)B_v\Delta v d\Omega$$

To find the total power in the frequency range Δv the antenna receives, this is integrated over all directions (a solid angle of 4π),

$$P_A = \frac{1}{2}\int_{4\pi} A_e(\theta,\phi)B_v\Delta v d\Omega$$

Since the antenna is isotropic, it has the same aperture $A_e(\theta,\phi) = A_e$ in any direction. So the aperture can be moved outside the integral. Similarly the radiance B_v in the cavity is the same in any direction:

$$P_A = \frac{1}{2}A_e B_v\Delta v \int_{4\pi} d\Omega$$

$$P_A = 2\pi A_e B_v\Delta v$$

Radio waves are low enough in frequency so the Rayleigh–Jeans formula gives a very close approximation of the blackbody spectral radiance:

$$B_v = \frac{2v^2 kT}{c^2} = \frac{2kT}{\lambda^2}$$

Therefore,

$$P_A = \frac{4\pi A_e kT}{\lambda^2}\Delta v$$

The Johnson–Nyquist noise power produced by a resistor at temperature T over a frequency range Δv is:

$$P_R = kT\Delta v$$

Since the cavities are in thermodynamic equilibrium $P_A = P_R$, so

$$\frac{4\pi A_e kT}{\lambda^2}\Delta v = kT\Delta v$$

$$A_e = \frac{\lambda^2}{4\pi}.$$

ANTENNA FACTOR

In electromagnetics, the antenna factor is defined as the ratio of the electric field strength to the voltage V (units: V or μV) induced across the terminals of an antenna. The voltage measured at the output terminals of an antenna is not the actual field intensity due to actual antenna gain, aperture characteristics, and loading effects.

For an electric field antenna, the field strength is in units of V/m or μV/m and the resulting antenna factor AF is in units of 1/m:

$$AF = \frac{E}{V}$$

If all quantities are expressed logarithmically in decibels instead of SI units, the above equation becomes,

$$AF_{dBm^{-1}} = E_{dBV/m} - V_{dBV} = E_{dB\mu V/m} - V_{dB\mu V}$$

For a magnetic field antenna, the field strength is in units of A/m and the resulting antenna factor is in units of A/(Vm). For the relationship between the electric and magnetic fields,

For a 50 Ω load, knowing that $P_D A_e = P_r = V^2/R$ and $E^2 = \sqrt{\dfrac{\mu_0}{\varepsilon_0}} P_D \sim 377 P_D$ (E and V noted here are the RMS values averaged over time), the antenna factor is developed as:

$$AF = \frac{\sqrt{377 P_D}}{\sqrt{50 P_D A_e}} = \frac{2.75}{\sqrt{A_e}} = \frac{9.73}{\lambda \sqrt{G}}$$

Where,

- $A_e = (\lambda^2 G)/4\pi$: the antenna effective aperture,
- P_D is the power density in watts per unit area,
- P_r is the power delivered into the load resistance presented by the receiver (normally 50 ohms),
- G: the antenna gain,
- μ_0 is the magnetic constant,
- ε_0 is the electric constant.

For antennas which are not defined by a physical area, such as monopoles and dipoles consisting of thin rod conductors, the effective length is used to measure the ratio between E and V.

ANTENNA GAIN

In electromagnetics, an antenna's power gain or simply gain is a key performance number which combines the antenna's directivity and electrical efficiency. In a transmitting antenna, the gain describes how well the antenna converts input power into radio waves headed in a specified direction. In a receiving antenna, the gain describes how well the antenna converts radio waves arriving from a specified direction into electrical power. When no direction is specified, "gain" is understood to refer to the peak value of the gain, the gain in the direction of the antenna's main lobe. A plot of the gain as a function of direction is called the gain pattern or radiation pattern.

Antenna gain is usually defined as the ratio of the power produced by the antenna from a far-field source on the antenna's beam axis to the power produced by a hypothetical lossless isotropic antenna, which is equally sensitive to signals from all directions. Usually this ratio is expressed in decibels, and these units are referred to as "decibels-isotropic" (dBi). An alternative definition compares the received power to the power received by a lossless half-wave dipole antenna, in which case the units are written as *dBd*. Since a lossless dipole antenna has a gain of 2.15 dBi, the relation between these units is $\text{Gain(dBd)} = \text{Gain(dBi)} - 2.15$. For a given frequency, the antenna's effective area is proportional to the power gain. An antenna's effective length is proportional to the *square root* of the antenna's gain for a particular frequency and radiation resistance. Due to reciprocity, the gain of any reciprocal antenna when receiving is equal to its gain when transmitting.

Directive gain or directivity is a different measure which does *not* take an antenna's electrical efficiency into account. This term is sometimes more relevant in the case of a receiving antenna where one is concerned mainly with the ability of an antenna to receive signals from one direction while rejecting interfering signals coming from a different direction.

Power Gain

Power gain (or simply gain) is a unitless measure that combines an antenna's efficiency $\epsilon_{antenna}$ and directivity D:

$$G = \epsilon_{antenna} \cdot D.$$

The notions of efficiency and directivity depend on the following.

Efficiency

A transmitting antenna accepts input power P_{in} at some point along the feedline. The point is typically taken to be at the antenna (the *feedpoint*), thereby not counting power lost due to joule heating in the feedline and reflections back down the feedline. The efficiency $\epsilon_{antenna}$ of an antenna is the total radiated power P_o divided by the input power at the feedpoint:

$$\epsilon_{antenna} = \frac{P_o}{P_{in}}$$

The electromagnetic reciprocity theorem guarantees that the electrical properties of an antenna, such as efficiency, directivity, and gain, are the same when the antenna is used for receiving as when it is transmitting.

Directivity

Antennas are invariably directional to a greater or lesser extent, according to how the output power is distributed in any given direction in three dimensions. We shall specify direction here in spherical coordinates (θ, ϕ), where θ is the altitude or angle above a specified reference plane (such as the ground), while ϕ is the azimuth as the angle between the projection of the given direction onto the reference plane and a specified reference direction (such as north or east) in that plane with specified sign (either clockwise or counterclockwise).

The distribution of output power as a function of the possible directions (θ, ϕ) is given by its radiation intensity $U(\theta, \phi)$ (in SI units: watts per steradian, W·sr^{-1}). The output power is obtained from the radiation intensity by integrating the latter over all solid angles $d\Omega = \sin\theta d\theta d\phi$:

$$P_o = \int_{-\pi}^{\pi}\int_{-\pi/2}^{\pi/2} U(\theta, \phi) d\Omega = \int_{-\pi}^{\pi}\int_{-\pi/2}^{\pi/2} U(\theta, \phi)\sin\theta d\theta d\phi.$$

The mean radiation intensity $\bar{U}$ is therefore given by:

$$\bar{U} = \frac{P_o}{4\pi} \quad \text{since there are 4π steradians in a sphere}$$

$$= \frac{\epsilon_{antenna} \cdot P_{in}}{4\pi} \quad \text{using the first formula for } P_o.$$

The directive gain or directivity $D(\theta, \phi)$ of an antenna in a given direction is the ratio of its radiation intensity $U(\theta, \phi)$ in that direction to its mean radiation intensity $\bar{U}$. That is,

$$D(\theta, \phi) = \frac{U(\theta, \phi)}{\bar{U}}.$$

An isotropic antenna, meaning one with the same radiation intensity in all directions, therefore has directivity 1 in all directions independently of its efficiency. More generally the maximum, minimum, and mean directivities of any antenna are always at least 1, at most 1, and exactly 1. For the half-wave dipole the respective values are 1.64 (2.15 dB), 0, and 1.

When the directivity D of an antenna is given independently of direction it refers to its maximum directivity in any direction, namely:

$$D = \max_{\theta, \phi} D(\theta, \phi).$$

Gain

The power gain or simply gain $G(\theta, \phi)$ of an antenna in a given direction takes efficiency into account by being defined as the ratio of its radiation intensity $U(\theta, \phi)$ in that direction to the mean radiation intensity of a perfectly efficient antenna. Since the latter equals $P_{in}/4\pi$, it is therefore given by,

$$G(\theta, \phi) = \frac{U(\theta, \phi)}{P_{in}/4\pi}$$

$$= \epsilon_{antenna} \cdot \frac{U(\theta, \phi)}{\bar{U}} \quad \text{using the second equation for } \bar{U}$$

$$= \epsilon_{antenna} \cdot D(\theta, \phi) \quad \text{using the equation for } D(\theta, \phi)$$

As with directivity, when the gain G of an antenna is given independently of direction it refers to its maximum gain in any direction. Since the only difference between gain and directivity in any direction is a constant factor of $\epsilon_{antenna}$ independent of θ and ϕ, we obtain the fundamental formula:

$$G = \epsilon_{antenna} \cdot D.$$

If only a certain portion of the electrical power received from the transmitter is actually radiated by the antenna (i.e. less than 100% efficiency), then the directive gain compares the power radiated in a given direction to that reduced power (instead of the total power received), ignoring the inefficiency. The directivity is therefore the maximum directive gain when taken over all directions, and is always *at least* 1. On the other hand, the power gain takes into account the poorer efficiency by comparing the radiated power in a given direction to the actual power that the antenna receives from the transmitter, which makes it a more useful figure of merit for the antenna's contribution to the ability of a transmitter in sending a radio wave toward a receiver. In every direction, the power gain of an isotropic antenna is equal to the efficiency, and hence is always *at most* 1, though it can and ideally should exceed 1 for a directional antenna.

Note that in the case of an impedance mismatch, P_{in} would be computed as the transmission line's incident power minus reflected power. Or equivalently, in terms of the rms voltage V at the antenna terminals:

$$P_{in} = V^2 \cdot Re\left\{\frac{1}{Z_{in}}\right\}$$

where Z_{in} is the feedpoint impedance.

Gain in Decibels

Published numbers for antenna gain are almost always expressed in decibels (dB), a logarithmic scale. From the gain factor G, one finds the gain in decibels as:

$$G_{dBi} = 10 \cdot \log_{10}(G).$$

Therefore, an antenna with a peak power gain of 5 would be said to have a gain of 7 dBi. "dBi" is used rather than just "dB" to emphasize that this is the gain according to the basic definition, in which the antenna is compared to an isotropic radiator.

When actual measurements of an antenna's gain are made by a laboratory, the field strength of the test antenna is measured when supplied with, say, 1 watt of transmitter power, at a certain distance. That field strength is compared to the field strength found using a so-called *reference antenna* at the same distance receiving the same power in order to determine the gain of the antenna under test. That ratio would be equal to G if the reference antenna were an isotropic radiator(irad).

However a true isotropic radiator cannot be built, so in practice a different antenna is used. This will often be a half-wave dipole, a very well understood and repeatable antenna that can be easily built for any frequency. The directive gain of a half-wave dipole is known to be 1.64 and it can be made nearly

100% efficient. Since the gain has been measured with respect to this reference antenna, the difference in the gain of the test antenna is often compared to that of the dipole. The "gain relative to a dipole" is thus often quoted and is denoted using "dBd" instead of "dBi" to avoid confusion. Therefore, in terms of the true gain (relative to an isotropic radiator) G, this figure for the gain is given by:

$$G_{dBd} = 10 \cdot \log_{10}\left(\frac{G}{1.64}\right).$$

For instance, the above antenna with a gain G=5 would have a gain with respect to a dipole of 5/1.64 = 3.05, or in decibels one would call this 10 log(3.05) = 4.84 dBd. In general:

$$G_{dBd} = G_{dBi} - 2.15 dB$$

Both dBi and dBd are in common use. When an antenna's maximum gain is specified in decibels (for instance, by a manufacturer) one must be certain as to whether this means the gain relative to an isotropic radiator or with respect to a dipole. If it specifies "dBi" or "dBd" then there is no ambiguity, but if only "dB" is specified then the fine print must be consulted. Either figure can be easily converted into the other using the above relationship.

Note that when considering an antenna's directional pattern, "gain with respect to a dipole" does *not* imply a comparison of that antenna's gain in each direction to a dipole's gain in that direction. Rather, it is a comparison between the antenna's gain in each direction to the *peak* gain of the dipole (1.64). In any direction, therefore, such numbers are 2.15 dB smaller than the gain expressed in dBi.

Partial Gain

Partial gain is calculated as power gain, but for a particular polarization. It is defined as the part of the radiation intensity U corresponding to a given polarization, divided by the total radiation intensity of an isotropic antenna.

$$G_\theta = 4\pi\left(\frac{U_\theta}{P_{in}}\right)$$

$$G_\phi = 4\pi\left(\frac{U_\phi}{P_{in}}\right)$$

where U_θ and U_ϕ represent the radiation intensity in a given direction contained in their respective E field component.

As a result of this definition, we can conclude that the total gain of an antenna is the sum of partial gains for any two orthogonal polarizations.

$$G = G_\theta + G_\phi.$$

Example Calculation

Suppose a lossless antenna has a radiation pattern given by:

$$U = B_0 \sin^3(\theta).$$

Let us find the gain of such an antenna.

Solution:

First we find the peak radiation intensity of this antenna:

$$U_{max} = B_0$$

The total radiated power can be found by integrating over all directions:

$$P_{rad} = \int_0^{2\pi} \int_0^\pi U(\theta,\phi)\sin(\theta)d\theta d\phi = 2\pi B_0 \int_0^\pi \sin^4(\theta)d\theta = B_0\left(\frac{3\pi^2}{4}\right)$$

$$D = 4\pi\left(\frac{U_{max}}{P_{rad}}\right) = 4\pi\left[\frac{B_0}{B_0\left(\dfrac{3\pi^2}{4}\right)}\right] = \frac{16}{3\pi} = 1.698$$

Since the antenna is specified as being lossless the radiation efficiency is 1. The maximum gain is then equal to:

$$G = \epsilon_{antenna} D = (1)(1.698) = 1.698.$$

$$G_{dBi} = 10\log_{10}(1.698) = 2.30 \text{ dBi}$$

Expressed relative to the gain of a half-wave dipole we would find:

$$G_{dBd} = 10\log_{10}(1.698/1.64) = 0.15 \text{ dBd}.$$

Realized Gain

Realized gain differs from the above definitions of gain in that it is "reduced by the losses due to the mismatch of the antenna input impedance to a specified impedance." This mismatch induces losses above the dissipative losses described above; therefore, Realized Gain will always be less than Gain.

Gain may be expressed as absolute gain if further clarification is required to differentiate it from realized gain.

Total Radiated Power

Total radiated power is the sum of all RF power radiated by the antenna when the source power is included in the measurement. TRP is expressed in Watts, or equivalent logarithmic expressions, often dBm or dBW.

TRP can be measured while in the close proximity of power-absorbing losses such as the body and hand of the Mobile Device Under Test User.

The TRP can be used to determine Body Loss (BoL). The Body Loss is considered as the ratio of TRP measured in the presence of losses and TRP measured while in free space.

ANTENNA IMPEDANCE

Antenna impedance relates the voltage to the current at the input to the antenna.

Let's say an antenna has an impedance of 50 ohms. This means that if a sinusoidal voltage is applied at the antenna terminals with an amplitude of 1 Volt, then the current will have an amplitude of 1/50 = 0.02 Amps. Since the impedance is a real number, the voltage is in-phase with the current.

Alternatively, suppose the impedance is given by a complex number, say Z=50 + j*50 ohms.

Note that "j" is the square root of -1. Imaginary numbers are there to give phase information. If the impedance is entirely real [Z=50 + j*0], then the voltage and current are exactly in time-phase. If the impedance is entirely imaginary [Z=0 + j*50], then the voltage leads the current by 90 degrees in phase.

If Z=50 + j*50, then the impedance has a magnitude equal to:

$$\sqrt{50^2 + 50^2} = 70.71$$

The phase will be equal to:

$$\tan^{-1}\left(\frac{\text{Im}(Z)}{\text{Re}(Z)}\right) = 45^0$$

This means the phase of the current will lag the voltage by 45 degrees. That is, the current waveform is delayed relative to the voltage waveform. To spell it out, if the voltage (with frequency *f*) at the antenna terminals is given by,

$$V(t) = \cos(2\pi ft)$$

The electric current will then be equal to:

$$I(t) = \frac{1}{70.71}\cos\left(2\pi f - \frac{\pi}{180}.45\right)$$

Hence, antenna impedance is a simple concept. Impedance relates the voltage and current at the input to the antenna. The real part of the antenna impedance represents power that is either radiated away or absorbed within the antenna. The imaginary part of the impedance represents power that is stored in the near field of the antenna. This is non-radiated power. An antenna with a real input impedance (zero imaginary part) is said to be resonant. Note that the impedance of an antenna will vary with frequency.

Low Frequency

When we are dealing with low frequencies, the transmission line that connects the transmitter or receiver to the antenna is short. Short in antenna theory always means "relative to a wavelength". Hence, 5 meters could be short or very long, depending on what frequency we are operating at. At 60 Hz, the wavelength is about 3100 miles, so the transmission line can almost always be neglected. However, at 2 GHz, the wavelength is 15 cm, so the little length of line within your cell phone can often be considered a 'long line'. Basically, if the line length is less than a tenth of a wavelength, it is reasonably considered a short line.

Consider an antenna (which is represented as an impedance given by ZA) hooked up to a voltage source (of magnitude V) with source impedance given by ZS. The equivalent circuit of this is shown in figure.

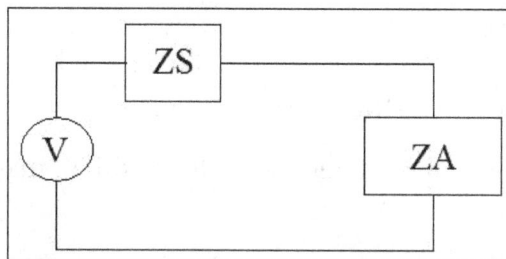

Circuit model of an antenna connected to a voltage source.

From circuit theory, we know that P=I*V. The power that is delivered to the antenna is:

$$P_A = \frac{V^2 \cdot ZA}{(ZA+ZS)^2}$$

If ZA is much smaller in magnitude than ZS, then no power will be delivered to the antenna and it won't transmit or receive energy. If ZA is much larger in magnitude than ZS, then no power will be delivered as well.

For maximum power to be transferred from the generator to the antenna, the ideal value for the antenna impedance is given by:

$$ZA + ZS^*$$

The * in the above equation represents complex conjugate. So if ZS=30+j*30 ohms, then for maximum power transfer the antenna should impedance ZA=30-j*30 ohms. Typically, the source impedance is real (imaginary part equals zero), in which case maximum power transfer occurs when ZA=ZS.

Hence, we now know that for an antenna to work properly, its impedance must not be too large or too small. It turns out that this is one of the fundamental design parameters for an antenna, and it isn't always easy to design an antenna with the right impedance - particularly over a wide frequency range.

High Frequency

In low-frequency circuit theory, the wires that connect things don't matter. Once the wires become a significant fraction of a wavelength, they make things very different. For instance, a short circuit has an impedance of zero ohms. However, if the impedance is measured at the end of a quarter wavelength transmission line, the impedance appears to be infinite, even though there is a dc conduction path.

In general, the transmission line will transform the impedance of an antenna, making it very difficult to deliver power, unless the antenna is matched to the transmission line. Consider the situation shown in figure. The impedance is to be measured at the end of a transmission line (with characteristic impedance Zo) and Length L. The end of the transmission line is hooked to an antenna with impedance ZA.

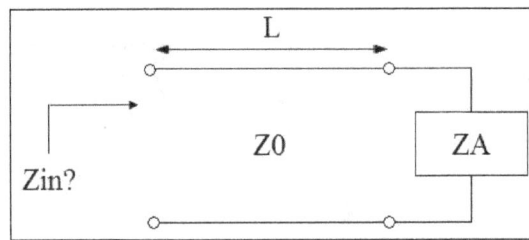

High Frequency Example.

It turns out (after studying transmission line theory for a while), that the input impedance Zin is given by:

$$Z_{in} = Z_0 \frac{ZA + jZ_0 \cdot \tan\left(\frac{2\pi f}{c} L\right)}{Z_0 + jZA \cdot \tan\left(\frac{2\pi f}{c} L\right)}$$

This is a little formidable for an equation to understand at a glance. However, the happy thing is If the antenna is matched to the transmission line (ZA=ZO), then the input impedance does not depend on the length of the transmission line.

This makes things much simpler. If the antenna is not matched, the input impedance will vary widely with the length of the transmission line. And if the input impedance isn't well matched to the source impedance, not very much power will be delivered to the antenna. This power ends up being reflected back to the generator, which can be a problem in itself (especially if high power is transmitted). This loss of power is known as *impedance mismatch*. Hence, we see that having a tuned impedance for an antenna is extremely important.

VSWR

We see that an antenna's impedance is important for minimizing impedance-mismatch loss. A poorly matched antenna will not radiate power. This can be somewhat alleviated via impedance matching, although this doesn't always work over a sufficient bandwidth.

A common measure of how well matched the antenna is to the transmission line or receiver is known as the Voltage Standing Wave Ratio (VSWR). VSWR is a real number that is always greater than or equal to 1. A VSWR of 1 indicates no mismatch loss (the antenna is perfectly matched to the tx line). Higher values of VSWR indicate more mismatch loss.

As an example of common VSWR values, a VSWR of 3.0 indicates about 75% of the power is delivered to the antenna (1.25 dB of mismatch loss); a VSWR of 7.0 indicates 44% of the power is delivered to the antenna (3.6 dB of mismatch loss). A VSWR of 6 or more is pretty high and will generally need to be improved.

The parameter VSWR sounds like an overly complicated concept; however, power reflected by an antenna on a transmission line interferes with the forward travelling power - and this creates a standing voltage wave - which can be numerically evaluated by the quantity Voltage Standing Wave Ratio (VSWR).

ANTENNA DIRECTIVITY

Directivity is a fundamental antenna parameter. It is a measure of how 'directional' an antenna's radiation pattern is. An antenna that radiates equally in all directions would have effectively zero directionality, and the directivity of this type of antenna would be 1 (or 0 dB).

An antenna's normalized radiation pattern can be written as a function in spherical coordinates:

$$F(\theta,\phi)$$

A normalized radiation pattern is the same as a radiation pattern; it is just scaled in magnitude such that the peak (maximum value) of the magnitude of the radiation pattern (F in equation below) is equal to 1. Mathematically, the formula for directivity (D) is written as:

$$D = \frac{1}{\dfrac{1}{4\pi} \int\limits_{0}^{2\pi}\int\limits_{0}^{\pi} \left|F(\theta,\phi)\right|^2 \sin\theta \, d\theta \, d\phi}$$

This equation for directivity might look complicated, but the numerator is the maximum value of F, and the denominator just represents the "average power radiated over all directions". This equation then is just a measure of the peak value of radiated power divided by the average, which gives the directivity of the antenna.

Directivity Example

As an example consider two antennas, one with radiation patterns given by:

$$F(\theta,\phi) = \sqrt{\sin(\theta)}$$

$$F(\theta,\phi) = (\sin\theta)^5$$

These radiation patterns are plotted in figure. Note that the patterns are only a function of the polar angle θ, and not a function of the azimuth angle (uniform in azimuth). The radiation pattern for antenna 1 is less directional then that for antenna 2; hence we expect the directivity to be lower.

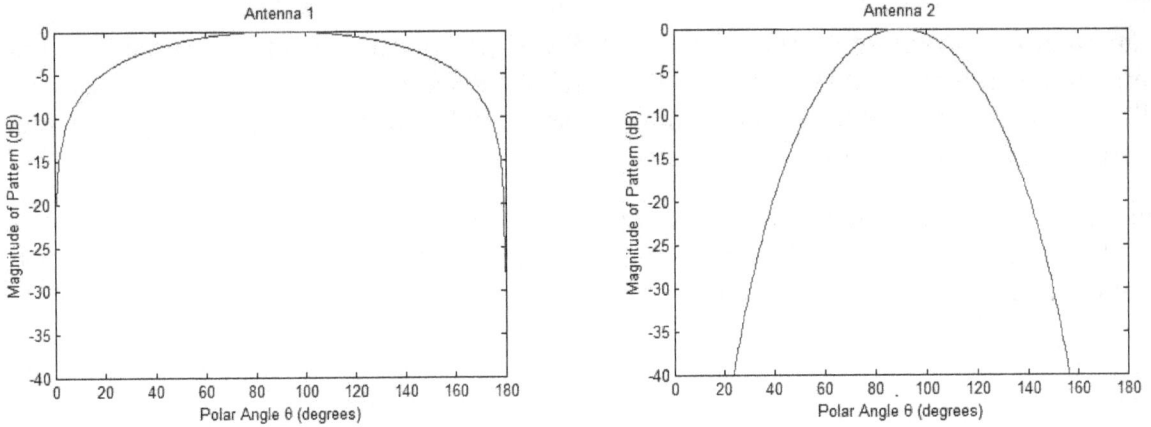

Plots of Radiation Patterns for Antennas. Which has the higher directivity?

Using Equation $F(\theta,\phi)$, we can figure out which antenna has the higher directivity. But to check your understanding, you should think about figure and what directivity is, and determine which has a higher directivity without using any mathematics.

The results of the directivity calculation, using equation:

- The directivity is calculated for Antenna 1 to be 1.273 (1.05 dB).

- The directivity is calculated for Antenna 2 to be 2.707 (4.32 dB).

Again, increased directivity implies a more 'focused' or 'directional' antenna. In words, Antenna 2 receives 2.707 times more power in its peak direction than an isotropic antenna would receive. Antenna 1 would receive 1.273 times the power of an isotropic antenna. The isotropic antenna is used as a common reference, even though no isotropic antennas exist.

Antennas for cell phones should have a low directivity because the signal can come from any direction, and the antenna should pick it up. In contrast, satellite dish antennas have a very high directivity, because they are to receive signals from a fixed direction. As an example, if you get a directTV dish, they will tell you where to point it such that the antenna will receive the signal.

Antenna Type	Typical Directivity	Typical Directivity (dB)
Short Dipole Antenna	1.5	1.76
Half-Wave Dipole Antenna	1.64	2.15
Patch (Microstrip) Antenna	3.2-6.3	5-8
Horn Antenna	10-100	10-20
Dish Antenna	10-10,000	10-40

As you can see from the above table, the directivity of an antenna can vary over several order of magnitude. Hence, it is important to understand directivity in choosing the best antenna for your specific application. If you need to transmit or receive energy from a wide variety of directions (example: car radio, mobile phones, computer wifi), then you should design an antenna with a low directivity. Conversely, if you are doing remote sensing, or targetted power transfer (example: received signal from a mountain top), you want a high directivity antenna, to maximize power transfer and reduce signal from unwanted directions.

Let's say we decide that we want an antenna with a low directivity. How do we accomplish this?

The general rule in Antenna Theory is that you need an electrically small antenna to produce low directivity. That is, if you use an antenna with a total size of 0.25 - 0.5 λ (a quarter- to a half-wavelength in size), then you will minimize directivity. That is, half-wave dipole antennas or half-wavelength slot antennas typically have directivities less than 3 dB, which is about as low of a directivity as you can obtain in practice. Ultimately, we can't make antennas much smaller than a quarter-wavelength without sacrificing antenna efficiency and antenna bandwidth.

Conversely, for antennas with a high directivity, we'll need antennas that are many wavelengths in size. That is, antennas such as dish (or satellite) antennas and horn antennas have high directivity, in part because they are many wavelengths long.

ANTENNA EFFICIENCY

The efficiency of an antenna is a ratio of the power delivered to the antenna relative to the power radiated from the antenna. A high efficiency antenna has most of the power present at the antenna's input radiated away. A low efficiency antenna has most of the power absorbed as losses within the antenna, or reflected away due to impedance mismatch.

One nice property of antennas is that the efficiency is the same whether we are using the antenna as a transmit or receive antenna. Hence, we could define antenna efficiency as the ratio of "potential power received from all possible angles", but that's more complicated. Just remember transmit and receive antenna efficiency is the same, and since it is easier to understand efficiency in terms of power radiated vs. power supplied, we'll simply use that definition. This property of antennas is known as antenna reciprocity.

The antenna efficiency (or radiation efficiency) can be written as the ratio of the radiated power to the input power of the antenna,

$$\varepsilon_R = \frac{P_{radiated}}{P_{input}}$$

Being a ratio, antenna efficiency is a number between 0 and 1. However, antenna efficiency is commonly quoted in terms of a percentage; for example, an efficiency of 0.5 is the same as 50%. Antenna efficiency is also frequently quoted in decibels (dB); an efficiency of 0.1 is 10% or (-10 dB), and an efficiency of 0.5 or 50% is -3 dB.

Equation above, is sometimes referred to as the antenna's radiation efficiency. This distinguishes it from another sometimes-used term, called an antenna's "total efficiency". The total efficiency of an antenna is the radiation efficiency multiplied by the impedance mismatch loss of the antenna, when connected to a transmission line or receiver (radio or transmitter). This can be summarized in equation below, where ε_T is the antenna's total efficiency, M_L is the antenna's loss due to impedance mismatch, and ε_R is the antenna's radiation efficiency.

$$\varepsilon_T = M_L \cdot \varepsilon_R$$

From equation above, since M_L is always a number between 0 and 1, the total antenna efficiency is always less than the antenna's radiation efficiency. Said another way, the radiation efficiency is the same as the total antenna efficiency if there was no loss due to impedance mismatch.

In practice, unless otherwise specified, antenna efficiency typically refers to the total efficiency (including mismatch loss).

What causes an antenna to not have an efficiency of 100% (or 0 dB)? Antenna efficiency losses are typically due to:

- Conduction losses (due to finite conductivity of the metal that forms the antenna).

- Dielectric losses (due to conductivity of a dielectric material near an antenna).

- Impedance mismatch loss.

Examples of dielectrics include glass, plastics, teflon, and rubber. The strong Electric Fields near an antenna lose energy to heat due to the conductivity of the dielectric. If the electrical conductivity is zero, the dielectric loss within a material is zero. However, many materials (such as silicone and glass) have conductivity that is low but still enough to significantly decrease the antenna efficiency.

Efficiency is one of the most important antenna parameters. It can be very close to 100% (or 0 dB) for dish antennas, horn antennas, or half-wavelength dipoles with no lossy materials around them. Mobile phone antennas, or wifi antennas in consumer electronics products, typically have efficiencies from 20%-70% (-7 to -1.5 dB). Car radio antennas can have an antenna efficiency of -20 dB (1% efficiency) at the AM radio frequencies; this is because the antennas are much smaller than a half-wavelength at the operational frequency, which greatly lowers antenna efficiency. The radio link is maintained because the AM Broadcast tower uses a very high transmit power.

It is very common in industry to quote antenna efficiency in percent. However, there are two strong reasons why antenna efficiency should be measured in decibels (dB):

- Everything associated with the RF (radio frequency) world is measured in dB: transmit power is dB, isolation is in dB, desense is in dB, radio sensitivity is in dB. Hence, it follows antenna efficiency should be in dB.

- If a change to an antenna is made, and someone says "how much did the efficiency change" and the response is "5%", that is ambiguous. An increase from 1% to 6% is a huge change (7.8 dB), whereas an increase from 85% to 90% is small (0.24 dB).

ANTENNA MEASUREMENT

Antenna measurement techniques refers to the testing of antennas to ensure that the antenna meets specifications or simply to characterize it. Typical parameters of antennas are gain, radiation pattern, beamwidth, polarization, and impedance.

The antenna pattern is the response of the antenna to a plane wave incident from a given direction or the relative power density of the wave transmitted by the antenna in a given direction. For a reciprocal antenna, these two patterns are identical. A multitude of antenna pattern measurement techniques have been developed. The first technique developed was the far-field range, where the antenna under test (AUT) is placed in the far-field of a range antenna. Due to the size required to create a far-field range for large antennas, near-field techniques were developed, which allow the measurement of the field on a surface close to the antenna (typically 3 to 10 times its wavelength). This measurement is then predicted to be the same at infinity. A third common method is the compact range, which uses a reflector to create a field near the AUT that looks approximately like a plane-wave.

Far-field Range (FF)

The far-field range was the original antenna measurement technique, and consists of placing the AUT a long distance away from the instrumentation antenna. Generally, the far-field distance or Fraunhofer distance, d, is considered to be,

$$d = \frac{2D^2}{\lambda},$$

where D is the maximum dimension of the antenna and λ is the wavelength of the radio wave. Separating the AUT and the instrumentation antenna by this distance reduces the phase variation across the AUT enough to obtain a reasonably good antenna pattern.

IEEE suggests the use of their antenna measurement standard, document number IEEE-Std-149-1979 for far-field ranges and measurement set-up for various techniques including ground-bounce type ranges.

Near-field Range (NF)

Planar Near-field Range

Planar near-field measurements are conducted by scanning a small probe antenna over a planar surface. These measurements are then transformed to the far-field by use of a Fourier transform, or more specifically by applying a method known as stationary phase to the Laplace transform . Three basic types of planar scans exist in near field measurements.

Rectangular Planar Scanning

The probe moves in the Cartesian coordinate system and its linear movement creates a regular rectangular sampling grid with a maximum near-field sample spacing of $\Delta x = \Delta y = \lambda / 2$.

Polar Planar Scanning

More complicated solution to the rectangular scanning method is the plane polar scanning method.

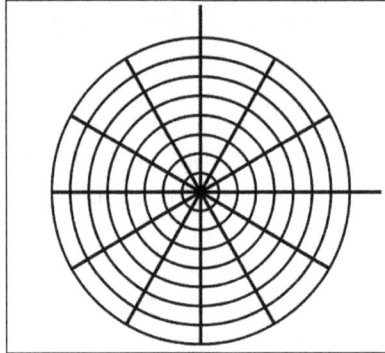

Bi-polar Planar Scanning

The bi-polar technique is very similar to the plane polar configuration.

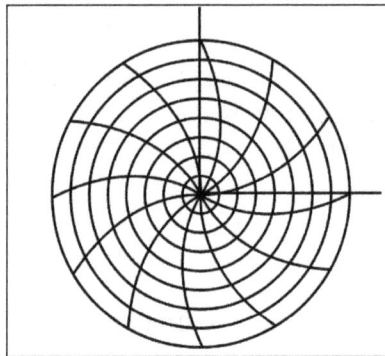

Cylindrical Near-field Range

Cylindrical near-field ranges measure the electric field on a cylindrical surface close to the AUT. Cylindrical harmonics are used transform these measurements to the far-field.

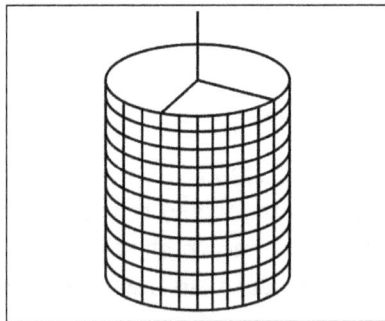

Spherical Near-field Range

Spherical near-field ranges measure the electric field on a spherical surface close to the AUT. Spherical harmonics are used transform these measurements to the far-field.

Free-space Ranges

The formula for electromagnetic radiation dispersion and information propagation is:

$$D^2 = \frac{P}{S} \propto 3dB$$

where D represents distance, P power and S speed.

The equation means that double the communication distance requires four times the power. It also means double power allows double communication speed (bit rate). Double power is approx. 3dB (10 log(2) to be exact) increase. Of course in the real world there are all sorts of other phenomena which enter in, such as Fresnel canceling, path loss, background noise, etc.

Compact Range

A Compact Antenna Test Range (CATR) is a facility which is used to provide convenient testing of antenna systems at frequencies where obtaining far-field spacing to the AUT would be infeasible using traditional free space methods. It was invented by Richard C. Johnson at the Georgia Tech Research Institute. The CATR uses a source antenna which radiates a spherical wavefront and one or more secondary reflectors to collimate the radiated spherical wavefront into a planar wavefront within the desired test zone. One typical embodiment uses a horn feed antenna and a parabolic reflector to accomplish this.

The CATR is used for microwave and millimeter wave frequencies where the $2\,D^2/\lambda$ far-field distance is large, such as with high-gain reflector antennas. The size of the range that is required can be much less than the size required for a full-size far-field anechoic chamber, although the cost of fabrication of the specially-designed CATR reflector can be expensive due to the need to ensure precision of the reflecting surface (typically less than λ/100 RMS surface accuracy) and to specially treat the edge of the reflector to avoid diffracted waves which can interfere with the desired beam pattern.

Elevated Range

A means of reducing reflection from waves bouncing off the ground.

Slant Range

A means of eliminating symmetrical wave reflection.

Physical Background

The electrical field created by an electric charge q is,

$$\vec{E} = \frac{-q}{4\pi\varepsilon_\circ}\left[\frac{\vec{e}_{r'}}{r'^2} + \frac{r'}{c}\frac{d}{dt}\left(\frac{\vec{e}_{r'}}{r'^2}\right) + \frac{1}{c^2}\frac{d^2}{dt^2}\left(\vec{e}_{r'}\right)\right]$$

where:

- c is the speed of light in vacuum.

- $\varepsilon_\circ$ is the permittivity of free space.

- r' is the distance from the observation point (the place where $\vec{E}$ is evaluated) to the point where the charge *was* $\dfrac{r'}{c}$ seconds before the time when the measure is done.

- $\vec{e}_{r'}$ is the unit vector directed from the observation point (the place where $\vec{E}$ is evaluated) to the point where the charge *was* $\dfrac{r'}{c}$ seconds before the time when the measure is done.

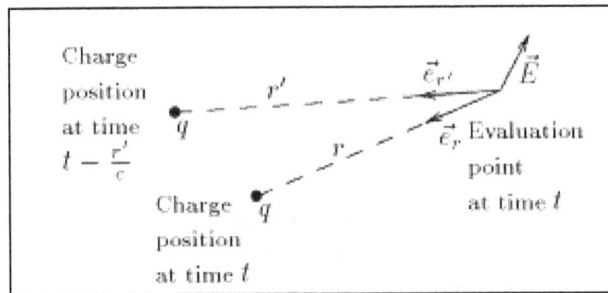

The measured electrical field was radiated $\dfrac{r'}{c}$ seconds earlier.

The "prime" in this formula appears because the electromagnetic signal travels at the speed of light. Signals are observed as coming from the point where they were emitted and not from the point where the emitter is at the time of observation. The stars that we see in the sky are no longer where we see them. We will see their current position years in the future; some of the stars that we see today no longer exist.

The first term in the formula is just the electrostatic field with retarded time.

The second term is *as though nature were trying to allow for the fact that the effect is retarded* (Feynman).

The third term is the only term that accounts for the far field of antennas.

The two first terms are proportional to $\dfrac{1}{r^2}$. Only the third is proportional to $\dfrac{1}{r}$.

Near the antenna, all the terms are important. However, if the distance is large enough, the first two terms become negligible and only the third remains:

$$\vec{E} = \frac{-q}{4\pi\varepsilon c_\circ^2}\frac{d^2}{dt^2}(\vec{e}_{r'}) = -q10^{-7}\frac{d^2}{dt^2}(\vec{e}_{r'})$$

If the charge q is in sinusoidal motion with amplitude $\ell_\circ$ and pulsation ω the power radiated by the charge is:

$$P = \frac{q^2\omega^4\ell_\circ^2}{12\pi\varepsilon_\circ c^3}\ \text{watts.}$$

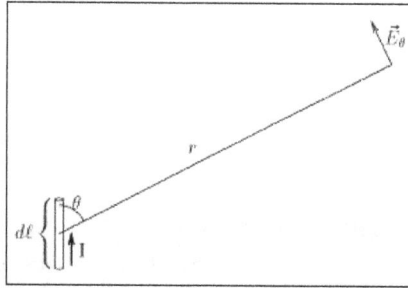

Electrical field radiated by an element of current. The element of current, the electrical field vector $\vec{E}_\theta$ and r are on the same plane.

Note that the radiated power is proportional to the fourth power of the frequency. It is far easier to radiate at high frequencies than at low frequencies. If the motion of charges is due to currents, it can be shown that the (small) electrical field radiated by a small length $d\ell$ of a conductor carrying a time varying current I is:

$$dE_\theta(t + \frac{r}{c}) = \frac{-d\ell \sin\theta}{4\pi\varepsilon_\circ c^2 r} \frac{dI}{dt}$$

The left side of this equation is the electrical field of the electromagnetic wave radiated by a small length of conductor. The index θ reminds that the field is perpendicular to the line to the source.

The $t + \frac{r}{c}$ reminds that this is the field observed $\frac{r}{c}$ seconds after the evaluation on the current derivative. The angle θ is the angle between the direction of the current and the direction to the point where the field is measured.

The electrical field and the radiated power are maximal in the plane perpendicular to the current element. They are zero in the direction of the current.

Only time-varying currents radiate electromagnetic power.

If the current is sinusoidal, it can be written in complex form, in the same way used for impedances. Only the real part is physically meaningful:

$$I = I_\circ e^{j\omega t}$$

where:

- $I_\circ$ is the amplitude of the current.

- $\omega = 2\pi f$ is the angular frequency.

- $j = \sqrt{-1}$.

The (small) electric field of the electromagnetic wave radiated by an element of current is:

$$dE_\theta(t + \frac{r}{c}) = \frac{-d\ell j\omega}{4\pi\varepsilon_\circ c^2} \frac{\sin\theta}{r} e^{j\omega t}$$

And for the time t:

$$dE_\theta(t) = \frac{-d\ell j\omega}{4\pi\varepsilon_\circ c^2} \frac{\sin\theta}{r} e^{j\left(\omega t - \frac{\omega}{c}r\right)}$$

The electric field of the electromagnetic wave radiated by an antenna formed by wires is the sum of all the electric fields radiated by all the small elements of current. This addition is complicated by the fact that the direction and phase of each of the electric fields are, in general, different.

Calculation of Antenna Parameters in Reception

The gain in any given direction and the impedance at a given frequency are the same when the antenna is used in transmission or in reception.

The electric field of an electromagnetic wave induces a small voltage in each small segment in all electric conductors. The induced voltage depends on the electrical field and the conductor length. The voltage depends also on the relative orientation of the segment and the electrical field.

Each small voltage induces a current and these currents circulate through a small part of the antenna impedance. The result of all those currents and tensions is far from immediate. However, using the reciprocity theorem, it is possible to prove that the Thévenin equivalent circuit of a receiving antenna is:

$$V_a = \frac{\sqrt{R_a G_a}\,\lambda\cos\psi}{2\sqrt{\pi Z_\circ}} E_b$$

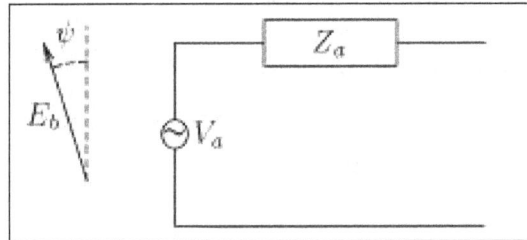

- V_a is the Thévenin equivalent circuit tension.

- Z_a is the Thévenin equivalent circuit impedance and is the same as the antenna impedance.

- R_a is the series resistive part of the antenna impedance Z_a.

- G_a is the directive gain of the antenna (the same as in emission) in the direction of arrival of electromagnetic waves.

- λ is the wavelength.

- E_b is the magnitude of the electrical field of the incoming electromagnetic wave.

- ψ is the angle of misalignment of the electrical field of the incoming wave with the antenna. For a dipole antenna, the maximum induced voltage is obtained when the electrical field is parallel to the dipole. If this is not the case and they are misaligned by an angle ψ, the induced voltage will be multiplied by $\cos\psi$.

- $Z_\circ = \sqrt{\dfrac{\mu_\circ}{\varepsilon_\circ}} = 376.730313461\,\Omega$ is a universal constant called vacuum impedance or impedance of free space.

The equivalent circuit and the formula at right are valid for any type of antenna. It can be as well a dipole antenna, a loop antenna, a parabolic antenna, or an antenna array.

From this formula, it is easy to prove the following definitions:

- Antenna effective length $= \dfrac{\sqrt{R_a G_a}\,\lambda \cos\psi}{\sqrt{\pi Z_\circ}}$

is the length which, multiplied by the electrical field of the received wave, give the voltage of the Thévenin equivalent antenna circuit.

- Maximum available power $= \dfrac{G_a \lambda^2}{4\pi Z_\circ} E_b^2$

is the maximum power that an antenna can extract from the incoming electromagnetic wave.

- Cross section or effective capture surface $= \dfrac{G_a}{4\pi}\lambda^2$,

is the surface which multiplied by the power per unit surface of the incoming wave, gives the maximum available power.

The maximum power that an antenna can extract from the electromagnetic field depends only on the gain of the antenna and the squared wavelength λ. It does not depend on the antenna dimensions.

Using the equivalent circuit, it can be shown that the maximum power is absorbed by the antenna when it is terminated with a load matched to the antenna input impedance. This also implies that under matched conditions, the amount of power re-radiated by the receiving antenna is equal to that absorbed.

RADIATION PATTERN

In the field of antenna design the term radiation pattern (or antenna pattern or far-field pattern) refers to the directional (angular) dependence of the strength of the radio waves from the antenna or other source.

Particularly in the fields of fiber optics, lasers, and integrated optics, the term radiation pattern may also be used as a synonym for the near-field pattern or Fresnel pattern. This refers to the positional dependence of the electromagnetic field in the near-field, or Fresnel region of the source. The near-field pattern is most commonly defined over a plane placed in front of the source, or over a cylindrical or spherical surface enclosing it.

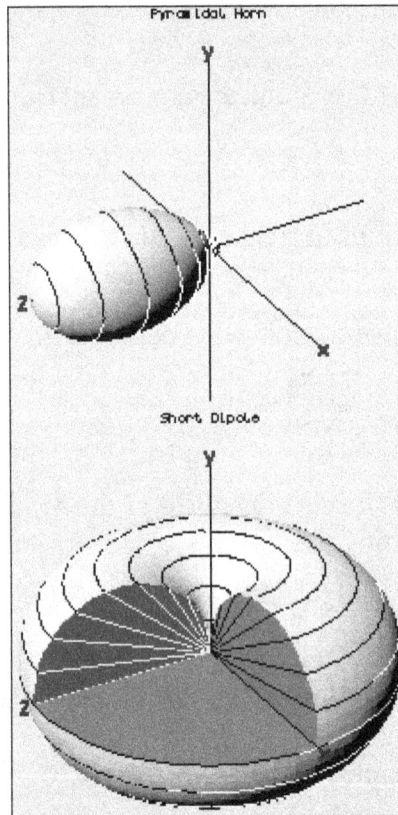

Three-dimensional antenna radiation patterns. The radial distance from the origin in any direction represents the strength of radiation emitted in that direction. The top shows the directive pattern of a horn antenna, the bottom shows the omnidirectional pattern of a simple vertical antenna.

The far-field pattern of an antenna may be determined experimentally at an antenna range, or alternatively, the near-field pattern may be found using a near-field scanner, and the radiation pattern deduced from it by computation. The far-field radiation pattern can also be calculated from the antenna shape by computer programs such as NEC. Other software, like HFSS can also compute the near field.

The far field radiation pattern may be represented graphically as a plot of one of a number of related variables, including; the field strength at a constant (large) radius (an amplitude pattern or field pattern), the power per unit solid angle (power pattern) and the directive gain. Very often, only the relative amplitude is plotted, normalized either to the amplitude on the antenna boresight, or to the total radiated power. The plotted quantity may be shown on a linear scale, or in dB. The plot is typically represented as a three-dimensional graph (as at right), or as separate graphs in the vertical plane and horizontal plane. This is often known as a polar diagram.

Reciprocity

It is a fundamental property of antennas that the receiving pattern (sensitivity as a function of direction) of an antenna when used for receiving is identical to the far-field radiation pattern of the antenna when used for transmitting. This is a consequence of the reciprocity theorem of electro-magnetics and is proved below. Therefore, in discussions of radiation patterns the antenna can be viewed as either transmitting or receiving, whichever is more convenient. Note however that this applies only to the passive antenna elements. Active antennas that include amplifiers or other components are no longer reciprocal devices.

Typical Patterns

Since electromagnetic radiation is dipole radiation, it is not possible to build an antenna that radiates coherently equally in all directions, although such a hypothetical isotropic antenna is used as a reference to calculate antenna gain.

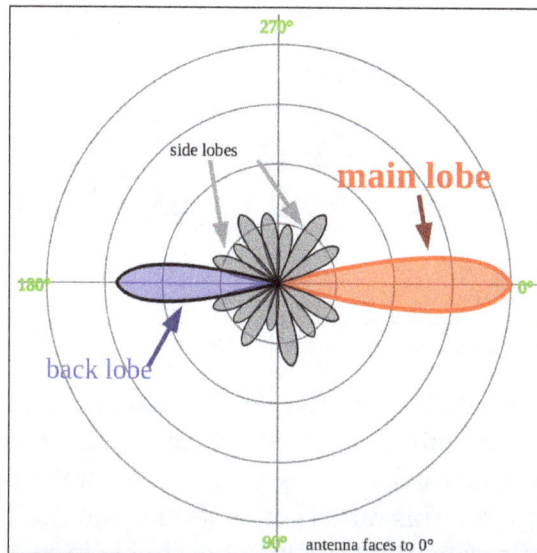

Typical polar radiation plot. Most antennas show a pattern of "lobes" or maxima of radiation. In a directive antenna, shown here, the largest lobe, in the desired direction of propagation, is called the "main lobe". The other lobes are called "sidelobes" and usually represent radiation in unwanted directions.

The simplest antennas, monopole and dipole antennas, consist of one or two straight metal rods along a common axis. These axially symmetric antennas have radiation patterns with a similar symmetry, called omnidirectional patterns; they radiate equal power in all directions perpendicular to the antenna, with the power varying only with the angle to the axis, dropping off to zero on the antenna's axis. This illustrates the general principle that if the shape of an antenna is symmetrical, its radiation pattern will have the same symmetry.

In most antennas, the radiation from the different parts of the antenna interferes at some angles. This results in zero radiation at certain angles where the radio waves from the different parts arrive out of phase, and local maxima of radiation at other angles where the radio waves arrive in phase. Therefore, the radiation plot of most antennas shows a pattern of maxima called "lobes" at various angles, separated by "nulls" at which the radiation goes to zero.

The larger the antenna is compared to a wavelength, the more lobes there will be. In a directive antenna in which the objective is to direct the radio waves in one particular direction, the lobe in that direction is larger than the others; this is called the "main lobe". The axis of maximum radiation, passing through the center of the main lobe, is called the "beam axis" or boresight axis". In some antennas, such as split-beam antennas, there may exist more than one major lobe. A minor lobe is any lobe except a major lobe.

The other lobes, representing unwanted radiation in other directions, are called "side lobes". The side lobe in the opposite direction (180°) from the main lobe is called the "back lobe".

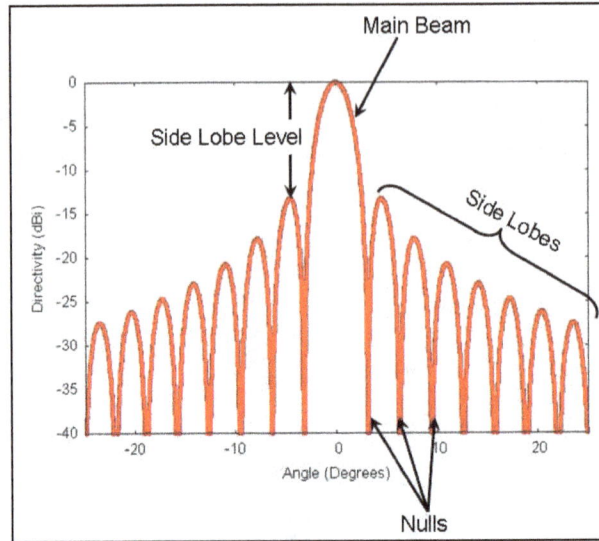

A rectangular radiation plot, an alternative presentation method to a polar plot.

Minor lobes usually represent radiation in undesired directions, so in directional antennas a design goal is usually to reduce the minor lobes. Side lobes are normally the largest of the minor lobes. The level of minor lobes is usually expressed as a ratio of the power density in the lobe in question to that of the major lobe. This ratio is often termed the side lobe ratio or side lobe level. Side lobe levels of −20 dB or greater are usually not desirable in many applications. Attainment of a side lobe level smaller than −30 dB usually requires very careful design and construction. In most radar systems, for example, low side lobe ratios are very important to minimize false target indications through the side lobes.

Proof of Reciprocity

We present a common simple proof limited to the approximation of two antennas separated by a large distance compared to the size of the antenna, in a homogeneous medium. The first antenna is the test antenna whose patterns are to be investigated; this antenna is free to point in any direction. The second antenna is a reference antenna, which points rigidly at the first antenna.

Each antenna is alternately connected to a transmitter having a particular source impedance, and a receiver having the same input impedance (the impedance may differ between the two antennas).

It is assumed that the two antennas are sufficiently far apart that the properties of the transmitting antenna are not affected by the load placed upon it by the receiving antenna. Consequently, the amount of power transferred from the transmitter to the receiver can be expressed as the product of two independent factors; one depending on the directional properties of the transmitting antenna, and the other depending on the directional properties of the receiving antenna.

For the transmitting antenna, by the definition of gain, G, the radiation power density at a distance r from the antenna (i.e. the power passing through unit area) is,

$$W(\theta,\Phi) = \frac{G(\theta,\Phi)}{4\pi r^2} P_t.$$

Here, the angles θ and Φ indicate a dependence on direction from the antenna, and P_t stands for the power the transmitter would deliver into a matched load. The gain G may be broken down into three factors; the antenna gain (the directional redistribution of the power), the radiation efficiency (accounting for ohmic losses in the antenna), and lastly the loss due to mismatch between the antenna and transmitter. Strictly, to include the mismatch, it should be called the realized gain, but this is not common usage.

For the receiving antenna, the power delivered to the receiver is:

$$P_r = A(\theta,\Phi)W.$$

Here W is the power density of the incident radiation, and A is the antenna aperture or effective area of the antenna (the area the antenna would need to occupy in order to intercept the observed captured power). The directional arguments are now relative to the receiving antenna, and again A is taken to include ohmic and mismatch losses.

Putting these expressions together, the power transferred from transmitter to receiver is:

$$P_r = A\frac{G}{4\pi r^2} P_t,$$

where G and A are directionally dependent properties of the transmitting and receiving antennas respectively.

For transmission from the reference antenna (2), to the test antenna (1), that is:

$$P_{1r} = A_1(\theta,\Phi)\frac{G_2}{4\pi r^2} P_{2t},$$

and for transmission in the opposite direction,

$$P_{2r} = A_2\frac{G_1(\theta,\Phi)}{4\pi r^2} P_{1t}.$$

Here, the gain G_2 and effective area A_2 of antenna 2 are fixed, because the orientation of this antenna is fixed with respect to the first.

Now for a given disposition of the antennas, the reciprocity theorem requires that the power transfer is equally effective in each direction, i.e.

$$\frac{P_{1r}}{P_{2t}} = \frac{P_{2r}}{P_{1t}},$$

whence ,

$$\frac{A_1(\theta, \Phi)}{G_1(\theta, \Phi)} = \frac{A_2}{G_2}.$$

But the right hand side of this equation is fixed (because the orientation of antenna 2 is fixed), and so,

$$\frac{A_1(\theta, \Phi)}{G_1(\theta, \Phi)} = \text{constant},$$

i.e. the directional dependence of the (receiving) effective aperture and the (transmitting) gain are identical (QED). Furthermore, the constant of proportionality is the same irrespective of the nature of the antenna, and so must be the same for all antennas. Analysis of a particular antenna (such as a Hertzian dipole), shows that this constant is $\frac{\lambda^2}{4\pi}$, where λ is the free-space wavelength.

Hence, for any antenna the gain and the effective aperture are related by,

$$A(\theta, \Phi) = \frac{\lambda^2 G(\theta, \Phi)}{4\pi}.$$

Even for a receiving antenna, it is more usual to state the gain than to specify the effective aperture. The power delivered to the receiver is therefore more usually written as:

$$P_r = \frac{\lambda^2 G_r G_t}{(4\pi r)^2} P_t$$

The effective aperture is however of interest for comparison with the actual physical size of the antenna.

Practical Consequences

- When determining the pattern of a receiving antenna by computer simulation, it is not necessary to perform a calculation for every possible angle of incidence. Instead, the radiation pattern of the antenna is determined by a single simulation, and the receiving pattern inferred by reciprocity.

- When determining the pattern of an antenna by measurement, the antenna may be either receiving or transmitting, whichever is more convenient.

- For a practical antenna, the side lobe level should be minimum, it is necessary to have the maximum directivity.

FIELD REGIONS

The near field and far field are regions of the electromagnetic field (EM) around an object, such as a transmitting antenna, or the result of radiation scattering off an object. Non-radiative 'near-field' behaviours of electromagnetic fields dominate close to the antenna or scattering object, while electromagnetic radiation 'far-field' behaviours dominate at greater distances.

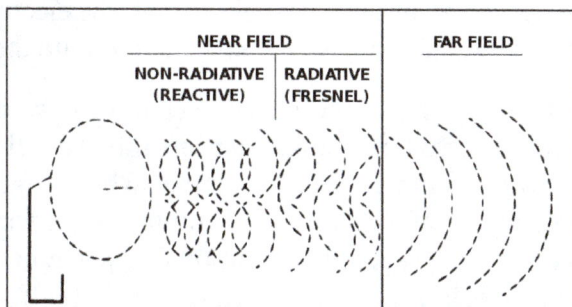

Differences between Fraunhofer diffraction and Fresnel diffraction.

Far-field E (electric) and B (magnetic) field strength decreases inversely with distance from the source, resulting in an inverse-square law for the radiated *power* intensity of electromagnetic radiation. By contrast, near-field E and B strength decrease more rapidly with distance: part decreases by the inverse-distance squared, the other part by an inverse cubed law, resulting in a diminished power in the parts of the electric field by an inverse fourth-power and sixth-power, respectively. The rapid drop in power contained in the near-field ensures that effects due to the near-field essentially vanish a few wavelengths away from the radiating part of the antenna.

Regions and their Interactions

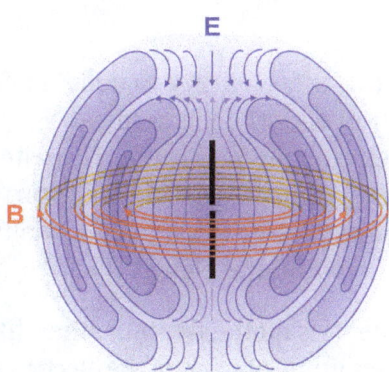

Near field: This dipole pattern shows a magnetic field $\vec{B}$ in red. The potential energy momentarily stored in this magnetic field is indicative of the reactive near field.

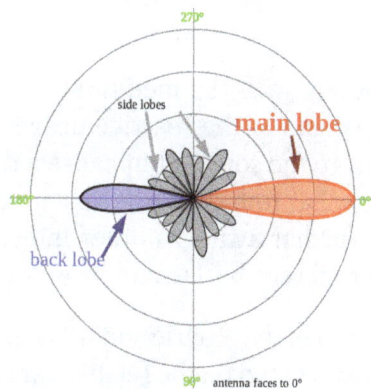

Far field: The radiation pattern can extend into the far field, where the reactive stored energy has no significant presence.

The far field is the region in which the field acts as "normal" electromagnetic radiation. In this region, it is dominated by electric or magnetic fields with electric dipole characteristics. The near field is governed by multipole type fields, which can be considered as collections of dipoles with a fixed phase relationship. The boundary between the two regions is only vaguely defined, and it depends on the dominant wavelength (λ) emitted by the source and the size of the radiating element.

In the far-field region of an antenna, radiated power decreases as the square of distance, and absorption of the radiation does not feed back to the transmitter. However, in the near-field region, absorption of radiation does affect the load on the transmitter. Magnetic induction as seen in a transformer can be seen as a very simple example of this type of near-field electromagnetic interaction.

In the far-field region, each part of the EM field (electric and magnetic) is "produced by" (or associated with) a change in the other part, and the ratio of electric and magnetic field intensities is simply the wave impedance. However, in the near-field region, the electric and magnetic fields can exist independently of each other, and one type of field can dominate the other.

In a normally-operating antenna, positive and negative charges have no way of leaving and are separated from each other by the excitation "signal" (a transmitter or other EM exciting potential). This generates an oscillating (or reversing) electrical dipole, which affects both the near field and the far field. In general, the purpose of antennas is to communicate wirelessly for long distances using far fields, and this is their main region of operation (however, certain antennas specialized for near-field communication do exist).

Also known as the radiation-zone field, the far field carries a relatively uniform wave pattern. The radiation zone is important because far fields in general fall off in amplitude by $1/r$. This means that the total energy per unit area at a distance r is proportional to $1/r^2$. The area of the sphere is proportional to r^2, so the total energy passing through the sphere is constant. This means that the far-field energy actually escapes to infinite distance (it *radiates*).

In contrast, the near field refers to regions such as near conductors and inside polarizable media where the propagation of electromagnetic waves is interfered with. One easy-to-observe example is the change of noise levels picked up by a set of rabbit ear antennas when one places a body part in close range. The near-field has been of increasing interest, particularly in the development of capacitive sensing technologies such as those used in the touchscreens of smart phones and tablet computers.

The interaction with the medium (e.g. body capacitance) can cause energy to deflect back to the source, as occurs in the reactive near field. Or the interaction with the medium can fail to return energy back to the source, but cause a distortion in the electromagnetic wave that deviates significantly from that found in free space, and this indicates the radiative near-field region, which is somewhat further away. Another intermediate region, called the transition zone, is defined on a somewhat different basis, namely antenna geometry and excitation wavelength.

The separation of the electric and magnetic fields into components is mathematical, rather than clearly physical, and is based on the relative rates that at which the amplitude of parts of the electric and magnetic fields diminish as distance from the radiating element increases. The amplitudes of the far-field components fall off as $1/r$, the radiative near-field amplitudes fall off as $1/r^2$, and the reactive near-field amplitudes fall off as $1/r^3$. Definitions of the regions attempt to characterize locations where the activity of the associated field components are the strongest. Mathematically, the distinction between field components is very clear, but the demarcation of the spatial field regions is subjective. All of the fields overlap everywhere, so for example, there are always substantial far-field and near-field radiative components in the closest-in near-field reactive region.

The regions defined below categorize field behaviors that are variable, even within the region of interest. Thus, the boundaries for these regions are approximate rules of thumb, as there are no precise cutoffs between them: All behavioral changes with distance are smooth changes. Even when precise boundaries can be defined in some cases, based primarily on antenna type and antenna size, experts may differ in their use of nomenclature to describe the regions. Because of these nuances, special care must be taken when interpreting technical literature that discusses "far-field" and "near-field" regions.

The term "near-field region" (also known as the "near field" or "near zone") has the following meanings with respect to different telecommunications technologies:

- The close-in region of an antenna where the angular field distribution is dependent upon the distance from the antenna.

- In the study of diffraction and antenna design, the near field is that part of the radiated field that is below distances shorter than the Fraunhofer distance, which is given by $d_F = \dfrac{2D^2}{\lambda}$ from the source of the diffracting edge or antenna of longitude or diameter D.

- In optical fiber communications, the region near a source or aperture that is closer than the Rayleigh length. (Presuming a Gaussian beam, which is appropriate for fiber optics).

Regions According to Electromagnetic Length

The most convenient practice is to define the size of the regions or zones in terms of fixed numbers (fractions) of wavelengths distant from the center of the radiating part of the antenna, with the clear understanding that the values chosen are only approximate and will be somewhat inappropriate for different antennas in different surroundings. The choice of the cut-off numbers is based on the relative strengths of the field component amplitudes typically seen in ordinary practice.

Electromagnetically Short Antennas

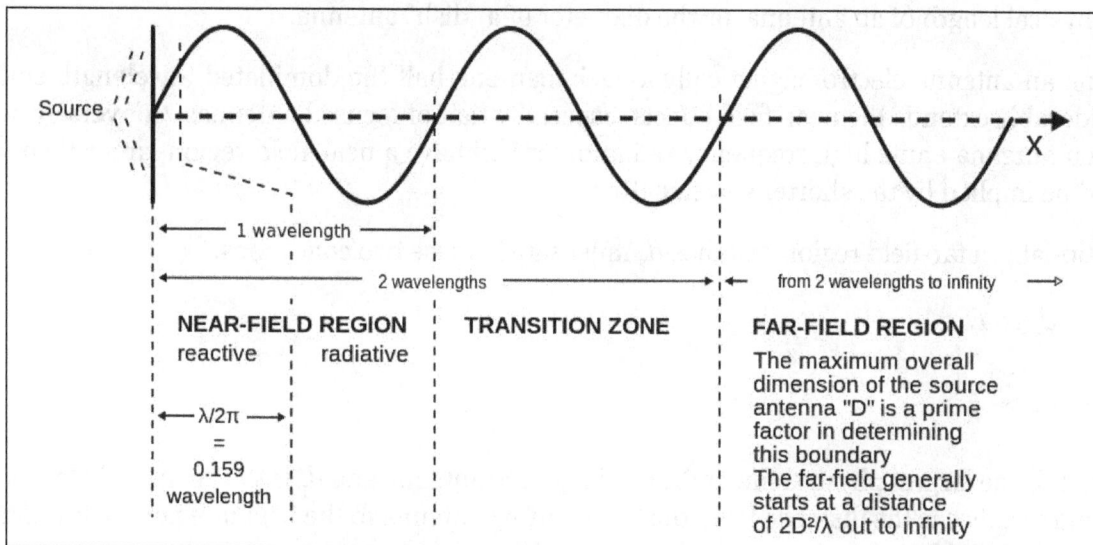

Field regions for antennas equal to, or shorter than, one-half wavelength of the radiation they emit, such as the whip antenna of a citizen's band radio, or an AM radio broadcast tower.

For antennas *shorter than half of the wavelength of the radiation they emit* (i.e., electromagnetically "short" antennas), the far and near regional boundaries are measured in terms of a simple ratio of the distance r from the radiating source to the wavelength λ of the radiation. For such an antenna, the near field is the region within a radius $r \ll \lambda$, while the far-field is the region for which $r \gg 2\lambda$. The transition zone is the region between $r = \lambda$ and $r = 2\lambda$.

Note that D, the length of the antenna is not important, and the approximation is the same for all shorter antennas (sometimes idealized as so-called "point antennas"). In all such antennas, the short length means that charges and currents in each sub-section of the antenna are the same at any given time, since the antenna is too short for the RF transmitter voltage to reverse before its effects on charges and currents are felt over the entire antenna length.

Electromagnetically Long Antennas

For antennas physically larger than a half-wavelength of the radiation they emit, the near and far fields are defined in terms of the Fraunhofer distance. Named after Joseph von Fraunhofer, the following formula gives the Fraunhofer distance:

$$d_F = \frac{2D^2}{\lambda}$$

where D is the largest dimension of the radiator (or the diameter of the antenna) and λ is the wavelength of the radio wave. Either of the following two relations are equivalent, emphasizing the size of the region in terms of wavelengths λ or diameters D:

$$d_F = 2\left(\frac{D}{\lambda}\right)^2 \lambda = 2\left(\frac{D}{\lambda}\right)D$$

This distance provides the limit between the near and far field. The parameter D corresponds to the physical length of an antenna, or the diameter of a "dish" antenna.

Having an antenna electromagnetically longer than one-half the dominated wavelength emitted considerably extends the near-field effects, especially that of focused antennas. Conversely, when a given antenna emits high frequency radiation, it will have a near-field region larger than what would be implied by the shorter wavelength.

Additionally, a far-field region distance d_F must satisfy these two conditions.

$$d_F \gg D$$

$$d_F \gg \lambda$$

where D is the largest physical linear dimension of the antenna and d_F is the far-field distance. The far-field distance is the distance from the transmitting antenna to the beginning of the Fraunhofer region, or far field.

Transition Zone

The "transition zone" between these near and far field regions, extending over the distance from one to two wavelengths from the antenna, is the intermediate region in which both near-field and far-field effects are important. In this region, near-field behavior dies out and ceases to be important, leaving far-field effects as dominant interactions.

Regions According to Diffraction Behavior

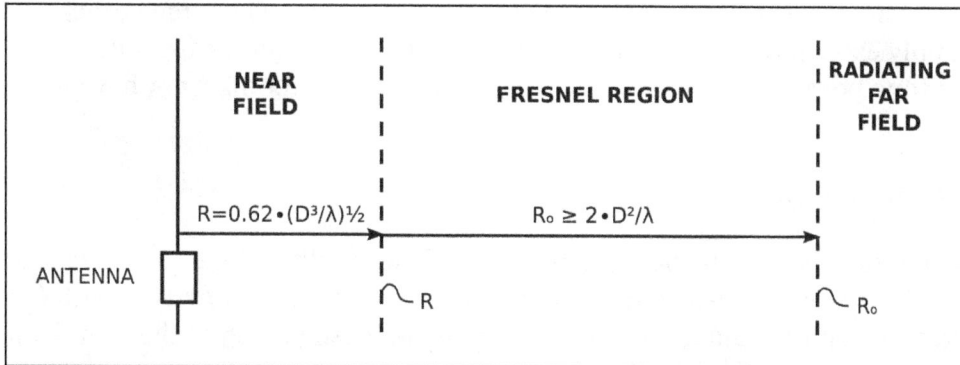

Near- and far-field regions for an antenna larger (diameter or length D) than the wavelength of the radiation it emits, so that $D/\lambda \gg 1$. Examples are radar dishes, satellite dish antennas, radio telescopes, and other highly directional antennas.

Far-field Diffraction

As far as acoustic wave sources are concerned, if the source has a maximum overall dimension or aperture width (D) that is large compared to the wavelength λ, the far-field region is commonly taken to exist at distances, when the Fresnel parameter S is larger than 1:

$$S = \frac{4\lambda}{D^2}r > 1, \text{ for } r > r_F = \frac{D^2}{4\lambda}.$$

For a beam focused at infinity, the far-field region is sometimes referred to as the "Fraunhofer region". Other synonyms are "far field", "far zone", and "radiation field". Any electromagnetic radiation consists of an electric field component E and a magnetic field component H. In the far field, the relationship between the electric field component E and the magnetic component H is that characteristic of any freely propagating wave, where E and H have equal magnitudes at any point in space (where measured in units where $c = 1$).

Near-field Diffraction

In contrast to the far field, the diffraction pattern in the near field typically differs significantly from that observed at infinity and varies with distance from the source. In the near field, the relationship between E and H becomes very complex. Also, unlike the far field where electromagnetic waves are usually characterized by a single polarization type (horizontal, vertical, circular, or elliptical), all four polarization types can be present in the near field.

The "near field" is a region in which there are strong inductive and capacitive effects from the currents and charges in the antenna that cause electromagnetic components that do not behave like far-field radiation. These effects decrease in power far more quickly with distance than do the far-field radiation effects. Non-propagating (or evanescent) fields extinguish very rapidly with distance, which makes their effects almost exclusively felt in the near-field region.

Also, in the part of the near field closest to the antenna (called the "reactive near field"), absorption of electromagnetic power in the region by a second device has effects that feed back to the transmitter, increasing the load on the transmitter that feeds the antenna by decreasing the antenna impedance that the transmitter "sees". Thus, the transmitter can sense when power is being absorbed in the closest near-field zone (by a second antenna or some other object) and is forced to supply extra power to its antenna, and to draw extra power from its own power supply, whereas if no power is being absorbed there, the transmitter does not have to supply extra power.

Near-field Characteristics

The near field itself is further divided into the *reactive* near field and the *radiative* near field. The "reactive" and "radiative" near-field designations are also a function of wavelength (or distance). However, these boundary regions are a fraction of one wavelength within the near field. The outer boundary of the reactive near-field region is commonly considered to be a distance of $\frac{1}{2\pi}$ times the wavelength $\frac{\lambda}{2\pi}$ or $0.159 \times \lambda$) from the antenna surface. The reactive near-field is also called the "inductive" near-field. The radiative near field (also called the "Fresnel region") covers the remainder of the near-field region, from $\frac{\lambda}{2\pi}$ out to the Fraunhofer distance.

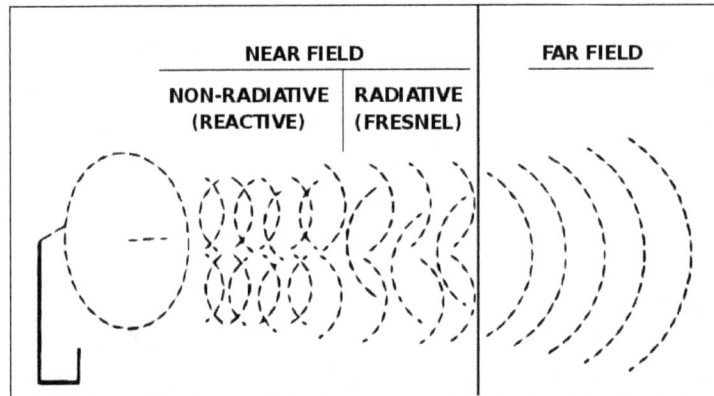

Differences between Fraunhofer diffraction and Fresnel diffraction.

Reactive Near Field or the Nearest Part of the Near Field

In the reactive near field (very close to the antenna), the relationship between the strengths of the E and H fields is often too complicated to easily predict, and difficult to measure. Either field component (E or H) may dominate at one point, and the opposite relationship dominate at a point only a short distance away. This makes finding the true power density in this region problematic. This is because to calculate power, not only E and H both have to be measured but the phase relationship between E and H as well as the angle between the two vectors must also be known in every point of space.

In this reactive region, not only is an electromagnetic wave being radiated outward into far space but there is a "reactive" component to the electromagnetic field, meaning that the nature of the field around the antenna is sensitive to EM absorption in this region, and reacts to it. In contrast, this is not true for absorption far from the antenna, which has no effect on the transmitter or antenna near field.

Very close to the antenna, in the reactive region, energy of a certain amount, if not absorbed by a receiver, is held back and is stored very near the antenna surface. This energy is carried back and forth from the antenna to the reactive near field by electromagnetic radiation of the type that slowly changes electrostatic and magnetostatic effects. For example, current flowing in the antenna creates a purely magnetic component in the near field, which then collapses as the antenna current begins to reverse, causing transfer of the field's magnetic energy back to electrons in the antenna as the changing magnetic field causes a self-inductive effect on the antenna that generated it. This returns energy to the antenna in a regenerative way, so that it is not lost. A similar process happens as electric charge builds up in one section of the antenna under the pressure of the signal voltage, and causes a local electric field around that section of antenna, due to the antenna's self-capacitance. When the signal reverses so that charge is allowed to flow away from this region again, the built-up electric field assists in pushing electrons back in the new direction of their flow, as with the discharge of any unipolar capacitor. This again transfers energy back to the antenna current.

Because of this energy storage and return effect, if either of the inductive or electrostatic effects in the reactive near field transfer any field energy to electrons in a different (nearby) conductor, then this energy is lost to the primary antenna. When this happens, an extra drain is seen on the transmitter, resulting from the reactive near-field energy that is not returned. This effect shows up as a different impedance in the antenna, as seen by the transmitter.

The reactive component of the near field can give ambiguous or undetermined results when attempting measurements in this region. In other regions, the power density is inversely proportional to the square of the distance from the antenna. In the vicinity very close to the antenna, however, the energy level can rise dramatically with only a small decrease in distance toward the antenna. This energy can adversely affect both humans and measurement equipment because of the high powers involved.

Radiative Near Field (Fresnel Region) or Farthest Part of the Near Field

The radiative near field (sometimes called the Fresnel region) does not contain reactive field components from the source antenna, since it is far enough from the antenna that back-coupling of the fields becomes out of phase with the antenna signal, and thus cannot efficiently return inductive or capacitive energy from antenna currents or charges. The energy in the radiative near field is thus all radiant energy, although its mixture of magnetic and electric components are still different from the far field. Further out into the radiative near field (one half wavelength to 1 wavelength from the source), the E and H field relationship is more predictable, but the E to H relationship is still complex. However, since the radiative near field is still part of the near field, there is potential for unanticipated (or adverse) conditions.

For example, metal objects such as steel beams can act as antennas by inductively receiving and then "re-radiating" some of the energy in the radiative near field, forming a new radiating surface to consider. Depending on antenna characteristics and frequencies, such coupling may be far more efficient than simple antenna reception in the yet-more-distant far field, so far more power may be transferred to the secondary "antenna" in this region than would be the case with a more distant antenna. When a secondary radiating antenna surface is thus activated, it then creates its own near-field regions, but the same conditions apply to them.

Compared to the Far Field

The near field is remarkable for reproducing classical electromagnetic induction and electric charge effects on the EM field, which effects "die-out" with increasing distance from the antenna: The magnetic field strength is proportional to the inverse-cube of the distance ($\frac{1}{r^3}$) and electric field strength proportional to inverse-square of distance ($\frac{1}{r^2}$). This fall-off is far more rapid than the classical *radiated* far-field (E and B fields, which are proportional to the simple inverse-distance ($\frac{1}{r}$). Typically near-field effects are not important farther away than a few wavelengths of the antenna.

More-distant near-field effects also involve energy transfer effects that couple directly to receivers near the antenna, affecting the power output of the transmitter if they do couple, but not otherwise. In a sense, the near field offers energy that is available to a receiver *only* if the energy is tapped, and this is sensed by the transmitter by means of responding to electromagnetic near fields emanating from the receiver. Again, this is the same principle that applies in induction coupled devices, such as a transformer, which draws more power at the primary circuit, if power is drawn from the secondary circuit. This is different with the far field, which constantly draws the same energy from the transmitter, whether it is immediately received, or not.

The amplitude of other components (non-radiative/non-dipole) of the electromagnetic field close to the antenna may be quite powerful, but, because of more rapid fall-off with distance than $\frac{1}{r}$ behavior, they do not radiate energy to infinite distances. Instead, their energies remain trapped in the region near the antenna, not drawing power from the transmitter unless they excite a receiver in the area close to the antenna. Thus, the near fields only transfer energy to very nearby receivers, and, when they do, the result is felt as an extra power draw in the transmitter. As an example of such an effect, power is transferred across space in a common transformer or metal detector by means of near-field phenomena (in this case inductive coupling), in a strictly "short-range" effect (i.e., the range within one wavelength of the signal).

Classical EM Modelling

Solving Maxwell's equations for the electric and magnetic fields for a localized oscillating source, such as an antenna, surrounded by a homogeneous material (typically vacuum or air), yields fields that, far away, decay in proportion to $\frac{1}{r}$ where r is the distance from the source. These are the *radiating* fields, and the region where r is large enough for these fields to dominate is the *far field*.

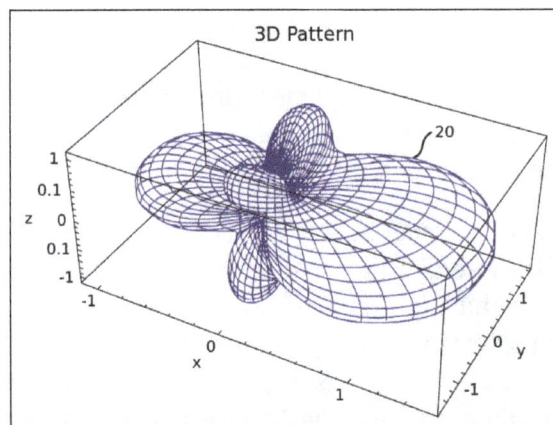

A "radiation pattern" for an antenna, by definition showing only the far field.

In general, the fields of a source in a homogeneous isotropic medium can be written as a multipole expansion. The terms in this expansion are spherical harmonics (which give the angular dependence) multiplied by spherical Bessel functions (which give the radial dependence). For large r, the spherical Bessel functions decay as $1/r$, giving the radiated field above. As one gets closer and closer to the source (smaller r), approaching the *near field*, other powers of r become significant.

The next term that becomes significant is proportional to $1/r^2$ and is sometimes called the *induction term*. It can be thought of as the primarily magnetic energy stored in the field, and returned to the antenna in every half-cycle, through self-induction. For even smaller r, terms proportional to $1/r^3$ become significant; this is sometimes called the *electrostatic field term* and can be thought of as stemming from the electrical charge in the antenna element.

Very close to the source, the multipole expansion is less useful (too many terms are required for an accurate description of the fields). Rather, in the near field, it is sometimes useful to express the contributions as a sum of radiating fields combined with evanescent fields, where the latter are exponentially decaying with r. And in the source itself, or as soon as one enters a region of inhomogeneous materials, the multipole expansion is no longer valid and the full solution of Maxwell's equations is generally required.

Antennas

If an oscillating electrical current is applied to a conductive structure of some type, electric and magnetic fields will appear in space about that structure. If those fields extend some distance into space the structure is often termed an antenna. Such an antenna can be an assemblage of conductors in space typical of radio devices or it can be an aperture with a given current distribution radiating into space as is typical of microwave or optical devices. The actual values of the fields in space about the antenna are usually quite complex and can vary with distance from the antenna in various ways.

However, in many practical applications, one is interested only in effects where the distance from the antenna to the observer is very much greater than the largest dimension of the transmitting antenna. The equations describing the fields created about the antenna can be simplified by assuming a large separation and dropping all terms that provide only minor contributions to the final field. These simplified distributions have been termed the "far field" and usually have the property that the angular distribution of energy does not change with distance, although the energy levels still vary with distance and time. Such an angular energy distribution is usually termed an antenna pattern.

Note that, by the principle of reciprocity, the pattern observed when a particular antenna is transmitting is identical to the pattern measured when the same antenna is used for reception. Typically one finds simple relations describing the antenna far-field patterns, often involving trigonometric functions or at worst Fourier or Hankel transform relationships between the antenna current distributions and the observed far-field patterns. While far-field simplifications are very useful in engineering calculations, this does not mean the near-field functions cannot be calculated, especially using modern computer techniques. An examination of how the near fields form about an antenna structure can give great insight into the operations of such devices.

Impedance

The electromagnetic field in the far-field region of an antenna is independent of the details of the near field and the nature of the antenna. The wave impedance is the ratio of the strength of the electric and magnetic fields, which in the far field are in phase with each other. Thus, the far field "impedance of free space" is resistive and is given by:

$$Z_0 \overset{\text{def}}{=} \mu_0 c_0 = \sqrt{\frac{\mu_0}{\varepsilon_0}} = \frac{1}{\varepsilon_0 c_0}.$$

With the usual approximation for the speed of light in free space $c_0 \approx 3.00 \times 10^8$ m/s, this gives the frequently used expression:

$$Z_0 = 119.92\pi\,\Omega \approx 120\pi\,\Omega \approx 377\Omega$$

The electromagnetic field in the near-field region of an electrically small coil antenna is predominantly magnetic. For small values of $\frac{r}{\lambda}$ the impedance of a magnetic loop is low and inductive, at short range being asymptotic to:

$$|Z_W| \approx 240\pi^2\,\frac{r}{\lambda} \approx 2370\,\frac{r}{\lambda}.$$

The electromagnetic field in the near-field region of an electrically short rod antenna is predominantly electric. For small values of $\frac{r}{\lambda}$ the impedance is high and capacitive, at short range being asymptotic to:

$$|Z_W| \approx 60\,\frac{\lambda}{r}.$$

In both cases, the wave impedance converges on that of free space as the range approaches the far field.

ANTENNA POLARISATION

Antenna polarisation (or polarization) is a very important consideration when choosing and installing an antenna and can mean as much as a 20db in signal loss if the receiver and transmitter antenna are not using the same polarisation. Most systems use either vertical, horizontal or circular polarisation.

Basic Field Equations

Suppose an electromagnetic wave, radiated by an antenna, has an electric field E (a vector) with two components: E_x and E_y.

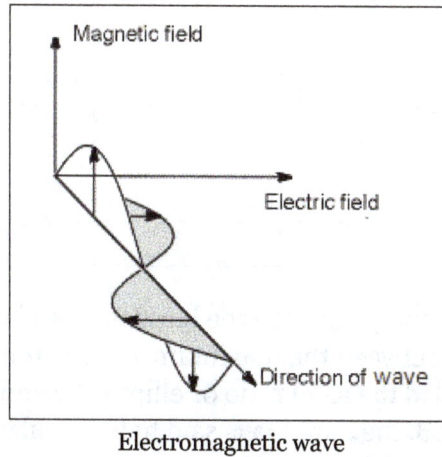

Electromagnetic wave

Let us assume that the components E_x and E_y of electric field E are given by,

$$E_x = a * \cos(\omega t - \beta z)$$
$$E_y = b * \cos(\omega t - \beta z + Phi)$$

where a is the amplitude of component E_x and b is the amplitude of component E. Phi is the difference of phase between the two components.

Polarisation

An antenna is a transducer that converts radio frequency electric current to electromagnetic waves that are then radiated into space. The electric field plane determines the polarisation or orientation of the radio wave. In general, most antennas radiate either linear or circular polarised.

A linear polarised antenna radiates wholly in one plane containing the direction of propagation. Where a circular polarised antenna, the plane of polarisation rotates in a circle making one complete revolution during one period of the wave. If the rotation is clockwise looking in the direction of propagation, the sense is called right-hand-circular (RHC). If the rotation is counter clockwise, the sense is called left-hand-circular (LHC).

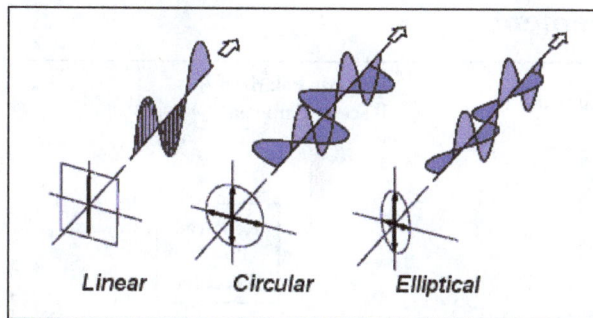

Linear   Circular   Elliptical

An antenna is said to be vertically polarised (linear) when its electric field is perpendicular to the Earth's surface. An example of a vertical antenna is a broadcast tower for AM radio or the "whip" antenna on an auto-mobile. Horizontally polarised (linear) antennas have their electric field parallel to the Earth's surface. Television transmissions use horizontal polarisation.

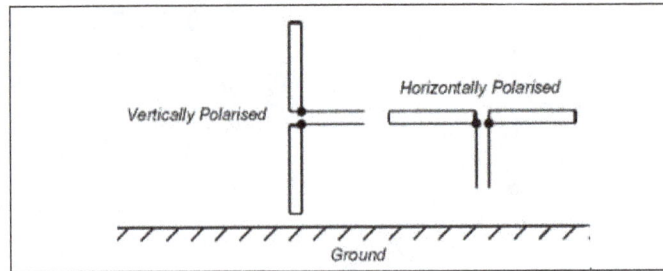

Circular polarised wave radiates energy in both the horizontal and vertical planes and all planes in between. The difference, if any, between the maximum and the minimum peaks as the antenna is rotated through all angles, is called the axial ratio or elliptically and is usually specified in decibels (dB). If the axial ratio is near 0 dB, the antenna is said to be circular polarised, when using a Helix Antenna. If the axial ratio is greater than 1-2 dB, the polarisation is often referred to as elliptical, when using a crossed Yagi.

Type of Polarization Required

Most communication systems use either vertical, horizontal or elliptical (RHC-right hand circular or LHC-left hand circular) polarization, with vertical dominating commercial VHF/UHF applications. In some instances, the selection is determined by the installation site, with the antenna oriented to provide the best performance. If this is anticipated, your antenna should provide mounting for either polarization.

Selecting the proper polarization for the system can enhance the overall performance by minimizing the interference from adjoining systems. For example, by installing you system orthogonal to other systems in the area, you can provide up to 20 dB of isolation. This will result in up to a 99% power reduction of the interfering system! Elliptical polarization can sometimes decrease fading.

Many systems are challenged because they must interface with handheld transmitters. These units move around a room or warehouse, with the antenna often pointing many degrees off-axis. To accommodate these application, the fixed antennas often use circular or elliptical polarization with a hemispherically shaped pattern trading off high gain for reasonable gain in all directions.

Polarisation Measurement

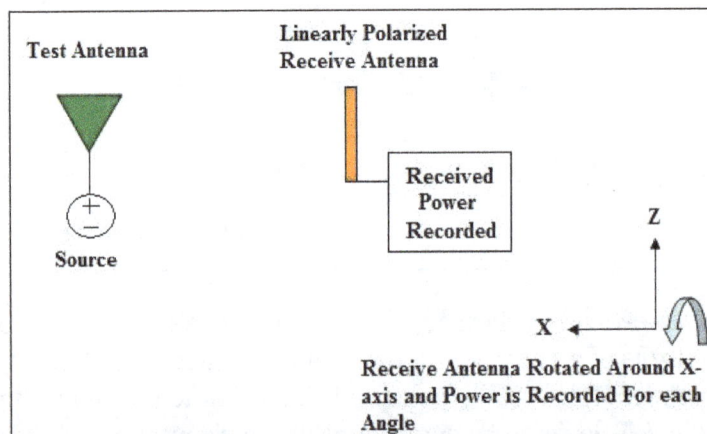

The polarization measurement method requires that a linearly polarized antenna, usually a dipole or a small horn, is rotated in the plane of polarization, which is taken to be normal to the direction of the incident field, and the output voltage of the probe is recorded. The recorded signal describes a polarization pattern for an elliptically polarized antenna. The polarization ellipse is tangent to the polarization pattern, and can be used to determine the axial ratio and the tilt angle of the AUT.

Considerations

Polarization is an important design consideration. The polarization of each antenna in a system should be properly aligned. Maximum signal strength between stations occurs when both stations are using identical polarization.

When choosing an antenna, it is an important consideration as to whether the polarization is linear or elliptical. If the polarization is linear, is it vertical or horizontal? If circular, is it RHC or LHC?

On line-of-sight (LOS) paths, it is most important that the polarization of the antennas at both ends of the path use the same polarization. In a linearly polarized system, a misalignment of polarization of 45 degrees will degrade the signal up to 3 dB and if misaligned 90 degrees the attenuation can be 20 dB or more. Likewise, in a circular polarized system, both antennas must have the same sense. If not, an additional loss of 20 dB or more will be incurred.

IMPEDANCE MATCHING

Impedance matching is the process of designing the antenna's input impedance (Z_{in}) or matching it to the corresponding RF circuitry's output impedance, which would be 50 Ω in most cases. A perfect match is obtained when $Z_L = Z_0$ ($Z_{in} = Z_0$) in equation $\Gamma = \dfrac{Z_L - Z_o}{Z_L + Z_o}$, which gives Γ a value of zero, and the SWR becomes unity in Equation 1. If the impedance of the line feeding the antenna and the antenna impedance do not match, then the source experiences complex impedance, which would be a function of the line length. Even if the antenna specifications say 50Ω impedance or matching is achieved using a matching network, the length of the line feeding the antenna is of significance, specifically if it is greater than approximately 1/10th the wavelength of the highest frequency of operation. Matching on the final board is crucial because antenna impedance can be altered depending on the electrical properties, size and proximity of the adjacent objects mounted on the end product, any enclosures, etc.

A Vector Network Analyzer (VNA) can be used to measure the input impedance of the antenna in the end-user environment, as this helps to optimize the antenna for the actual operating conditions. The VNA should be calibrated as close to the measurement plane as possible or at the location of the matching network. The impedance matching technique should consider any length of the transmission line if present between the calibration point and the matching network. A VNA can be used to measure S11 characteristics and to view the impedance on a Smith chart.

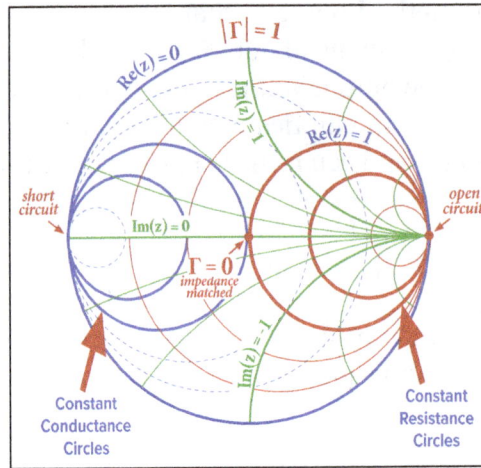

Typical Smith chart view.

A Smith chart is an excellent graphical aid for visualizing the impedance at any point of the transmission line or at the input of the antenna system across the different frequencies. A Smith chart consists of constant resistant circles and constant conductance circles as shown in figure.

A Smith chart can be used to perform an impedance match by bringing impedance to the center of the chart, which corresponds to a pure resistance of 50Ω by adjusting the reactance values. This is achieved by designing a matching network, which is a circuit between the feed line and the antenna. A Smith chart can be used to determine matching network's lumped element values.

Impedance Matching Methods

Antenna impedance is complex, consisting of both resistive and reactive parts, so the matching network must include components of both types to achieve optimal matching. If the source impedance is purely resistive and the load impedance is of complex type, then a complex conjugate of the load impedance would be required for the matching network.

In other words, for a load impedance of R+j*X, the impedance of the matching network would be R-j*X or the other way around. If the arbitrary impedance is at Point O, then the result of adding lumped elements in the matching network would be as shown in figure.

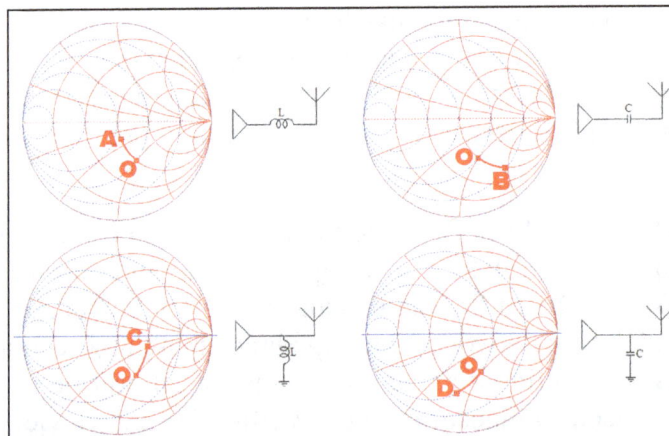

Impedance transformations for lumped network elements.

When a series inductor is connected to the antenna, the combined impedance of the antenna and the series inductor at the output will move toward Direction A in a constant resistant circle. A series capacitor will move the impedance toward Direction B along a constant resistant circle. A shunt inductor will move the impedance toward Direction C and a shunt capacitor will move the impedance toward Direction D along a constant conductance circle.

A few golden rules that simplify impedance matching are provided below.

1. A Smith chart can be divided into two halves: The upper half is inductive, and the lower half is capacitive.

2. Whenever impedance is to be moved up, an inductor (L) is required; use a capacitor (C) if the impedance is to be moved down.

3. The right/constant resistance circle is the series circle, and the left/constant conductance circle is the shunt circle. So, a shunt element is required if the impedance needs to be moved along the left circle. Otherwise, use a series element.

The addition of a series L or C will only match the impedances lying on the constant resistance circle, and a shunt L or C will match the impedances lying on the constant conductance circle. Multiple matching network element combinations can achieve the desired matching impedance. Other requirements such as filter type, Q-factor, and specific components can also be considered.

By combining any of the series inductor, series capacitor, the shunt inductor, and the shunt capacitor, any value of the load impedance in the Smith chart can be matched, except those spots located on the |Γ| = 1 circle, where the impedance is purely resistive.

Transmission lines are most commonly used to match the real impedances. Impedance matching at a frequency can be achieved by extending the length of the transmission line to bring the impedance on the Smith chart to reach the unit conductance circle that contains the Γ=0 point; then suitable shunt reactance is added to move the combined impedance to the Γ=0 point. The arbitrary impedance can also be rotated until it reaches the 50 Ω circle; then appropriate series reactance is added to get the resulting impedance to the 50 Ω point.

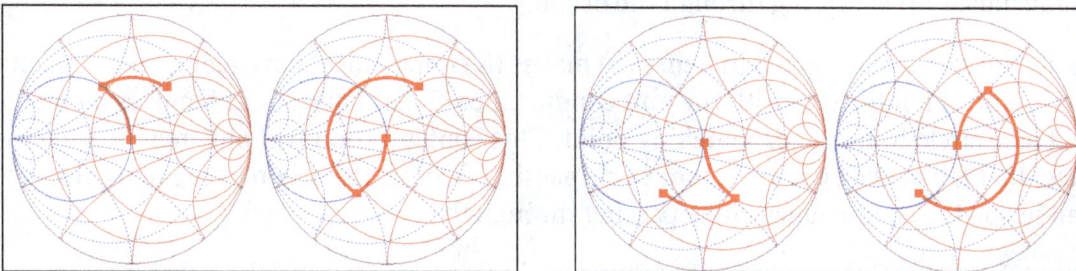

Any matching network can only move a limited portion of the impedance curve to the target matching circle on the Smith chart, which means there is a bandwidth limit for any matching network. The matching circuit can be used to achieve both impedance matching and bandwidth enhancement, and this is achieved by appropriately using the above principles while arranging the lumped elements in the form of L-networks, Pi-networks, and T-networks.

A simple L matching network consists of two lumped components, L and C, arranged in any of the eight different configurations shown in figure. Not every L-network configuration can guarantee the required matching between the given arbitrary load and source impedances. Each of these configurations has certain forbidden areas where matching cannot be achieved. Select the appropriate L-section topology based on where RL lies.

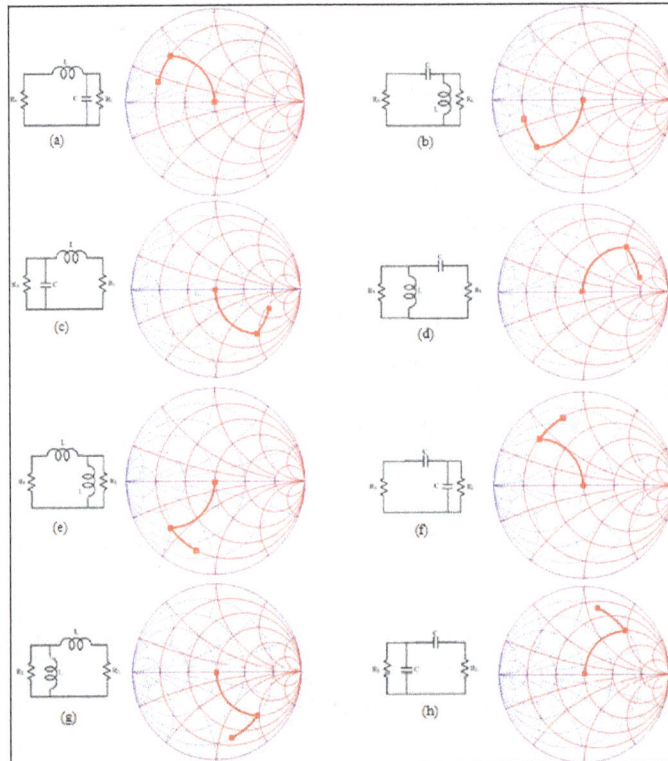

Impedance transformations for L-network matching topologies.

The series reactance L in figure (a) makes the impedance move along the constant resistance circle until it intersects with the unit conductance circle; then shunt C moves along the unit conductance circle to the 50 Ω matching point. This configuration will only match capacitive impedances that fall outside the resistance circle passing through the origin or inductive impedances that fall inside the conductance circle passing through the origin.

Similarly, the series reactance C in figure (b) makes the impedance move along the constant resistance circle until it intersects with the unit conductance circle. Then shunt L moves along the unit conductance circle to the 50 Ω matching point. This configuration will only match inductive impedances that fall outside the resistance circle passing through the origin or capacitive impedances that fall inside the conductance circle passing through the origin.

The shunt reactance C shown in figure (c) makes the impedance move along the constant conductance circle until it intersects with the unit resistant circle. Then the series L moves along the unit resistant circle to the 50 Ω matching point. This configuration will only match inductive impedances that fall outside the resistant circle passing through the origin or capacitive impedances that fall inside the resistant circle passing through the origin.

Similarly, the shunt reactance L in figure (d) makes the impedance move along the constant conductance circle until it intersects with the unit resistant circle; then series C moves along the unit resistant circle to the 50 Ω matching point. This configuration will only match capacitive impedances that fall outside the conductance circle passing through the origin or inductive impedances that fall inside the resistant circle passing through the origin.

An L-network with only inductive reactances can only match capacitive impedances that fall outside the resistant and conductance circles passing through the origin. An L-network with only capacitive reactances can only match inductive impedances that fall outside the resistant and conductance circles passing through the origin. The permissible and forbidden areas for the different types of L-networks from figure above are shown in figure below (I) and (II) respectively.

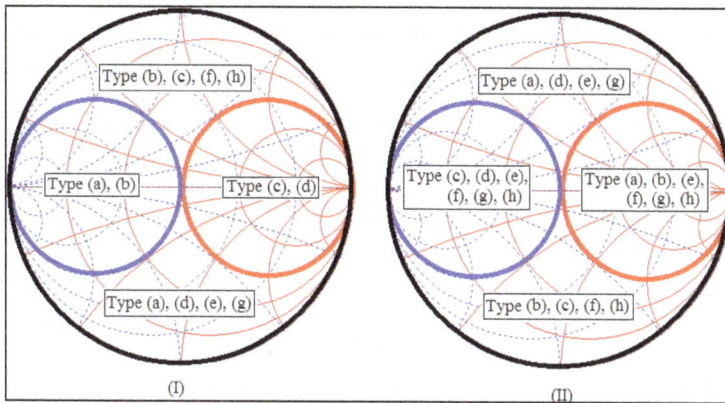

Permissible (I) and Forbidden (II) areas for L network matching

The matching circuit bandwidth depends on Q factor and frequency, where the Q factor of the circuit is based on source and load impedances, and the L-network does not provide considerable control over Q.

BW = F / Q

To enhance the bandwidth over which impedance is matched, an additional L-network can be included to the single section. In such scenarios, the impedance transforms to intermediate impedance using the first section. Then, the second L-network section matches the impedance to the desired value. If the bandwidth requirements are still not met, additional sections can be included to guarantee impedance matching over a wider bandwidth.

In such a network where there is more than one section of L-network, the termination ratios can be reduced, decreasing the Q of each section and thus enhancing the matched bandwidth of the circuit. Three-element networks, like Pi or T, provide greater flexibility to control the circuit Q, thereby controlling bandwidth while providing impedance matching.

Once the matching circuit is decided, the impedance can be measured using the builtin Smith chart function in a VNA for accuracy. Please take note that inaccurate calibration or port extension can give wrong results, which would need to be rectified and recalibrated for accuracy

Despite being a complicated procedure requiring testing and validation, antenna matching is essential in all RF designs. Impedance matching ensures maximum efficiency. Without proper matching, the antenna becomes a choke point of performance due to reduced range, increased power consumption and poor transmission quality.

References

- Antenna-theory-beam-width, antenna-theory: tutorialspoint.com, Retrieved 4 January 2019

- Stutzman, Warren L.; Thiele, Gary A. (2012). Antenna Theory and Design (3rd ed.). John Wiley & Sons. Pp. 560–564. ISBN 978-0470576649

- Condon, J. J.; Ransom, S. M. (2016). "Antenna Fundamentals". Essential Radio Astronomy course. US National Radio Astronomy Observatory (NRAO) website. Retrieved 22 August 2018

- Directivity, basics: antenna-theory.com , Retrieved 15 February 2019

- Jeffrey A. Nanzer, Microwave and Millimeter-wave Remote Sensing for Security Applications, pp. 268-269, Artech House, 2012 ISBN 1608071723

- Antenna-polarisation, learn-electronics, resources: electronicsforu.com, Retrieved 27, February 2019

- Impedance, basics: antenna-theory.com, Retrieved 9, January 2019

- Sundararajan, D. (4 March 2009). A Practical Approach to Signals and Systems. John Wiley & Sons. P. 109. ISBN 978-0-470-82354-5

3

Wire Antennas

Wire antenna is a form of antenna that consists of a long wire whose length does not depend on the wavelength of the radio waves. A few aspects that are studied in relation to wire antennas are transmission lines, short dipole, monopole, etc. The topics elaborated in this chapter will help in gaining a better perspective about wire antennas.

Wire antennas are the basic types of antennas. These are well known and widely used antennas. To have a better idea of these wire antennas, first let us have a look at the transmission lines.

Transmission Lines

The wire or the transmission line has some power, which travels from one end to the other end. If both the ends of transmission line are connected to circuits, then the information will be transmitted or received using this wire between these two circuits.

If one end of this wire is not connected, then the power in it tries to escape. This leads to wireless communication. If one end of the wire is bent, then the energy tries to escape from the transmission line, more effectively than before. This purposeful escape is known as Radiation.

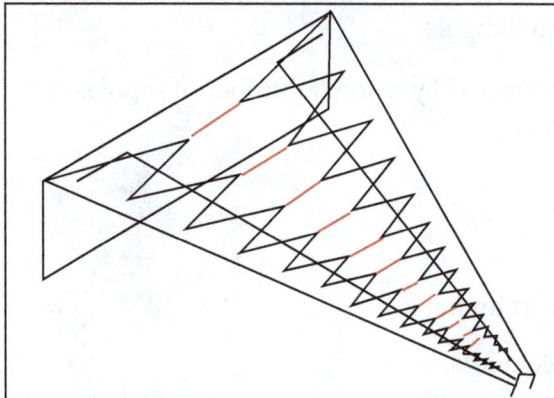

For the radiation to take place effectively, the impedance of the open end of the transmission line should match with the impedance of the free-space. Consider a transmission line of a quarter-wave length size. The far end of it is kept open and bent to provide high impedance. This acts as a half-wave dipole antenna. Already, it has low impedance at one end of the transmission line. The open end, which has high impedance, matches with the impedance of free space to provide better radiation.

Dipole

The radiation of energy when done through such a bent wire, the end of such transmission line is termed as dipole or dipole antenna.

The reactance of the input impedance is a function of the radius and length of the dipole. The smaller the radius, the larger the amplitude of the reactance. It is proportional to the wavelength. Hence, the length and radius of the dipole should also be taken into consideration. Normally, its impedance is around 72Ω.

This is better understood with the help of the following figure:

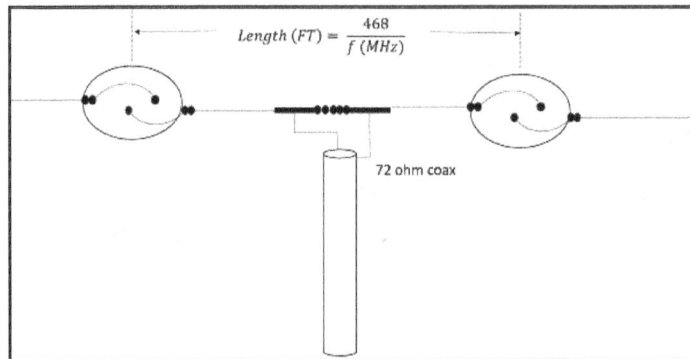

$$Length\ (FT) = \frac{468}{f\ (MHz)}$$

72 ohm coax

A dipole is a half-wave antenna fed at the center.
Here, the impedence at the center is near 72 ohms.

The figure shows the circuit diagram of a normal dipole connected to a transmission line. The current for a dipole is maximum at the center and minimum at its ends. The voltage is minimum at its center and maximum at its ends.

The types of wire antennas include Half-wave dipole, Half-wave folded dipole, Full-wave dipole, Short dipole, and Infinitesimal dipole.

The long wire antennas are formed by using a number of dipoles. The length of the wire in these type of antennas is n times $\lambda/2$,

$$L = n\lambda/2$$

Where,

- L is the length of the antenna,
- n is the number of elements,
- λ is the wavelength.

As 'n' increases, the directional properties also increase.

Types of Long-wire Antennas

Resonant Antennas

Resonant Antennas are those for which a sharp peak in the radiated power is intercepted by the antenna at certain frequency, to form a standing wave. The radiation pattern of the radiated wave is not matched with the load impedance in this type of antenna.

The resonant antennas are periodic in nature. They are also called as bi-directional travelling wave antennas, as the radiated wave moves in two directions, which means both incident and reflected waves occur here. In these antennas, the length of the antenna and frequency are proportional to each other.

Non-resonant Antennas

Non-resonant Antennas are those for which resonant frequency does not occur. The wave moves in forward direction and hence do not form a standing wave. The radiation pattern of the radiated wave matches with the load impedance in the non-resonant antennas.

These non-resonant antennas are non-periodic in nature. They are also called as Unidirectional travelling wave antennas, as the radiated wave moves in forward direction only, which means that only incident wave is present. As the frequency increases, the length of the antenna decreases and vice versa. Hence, the frequency and length are inversely proportional to each other.

These long-wire antennas are the basic elements for the construction of V-shaped antennas or the Rhombic antennas.

LOOP ANTENNA

A loop antenna is a radio antenna consisting of a loop or coil of wire, tubing, or other electrical conductor usually fed by a balanced source or feeding a balanced load. Within this physical description there are two distinct antenna types. The large self-resonant loop antenna has a circumference close to one wavelength of the operating frequency and so is resonant at that frequency. This category also includes smaller loops 5% to 30% of a wavelength in circumference, which use a capacitor to make them resonant. These antennas are used for both transmission and reception. In contrast, small loop antennas less than 1% of a wavelength in size are very inefficient radiators, and so are only used for reception. An example is the ferrite (loopstick) antenna used in most AM broadcast radios. Loop antennas have a dipole radiation pattern; they are most sensitive to radio waves in two broad lobes in opposite directions, 180° apart. Due to this directional pattern they are used for radio direction finding (RDF), to locate the position of a transmitter.

A shortwave loop antenna.

Full-size (Self Resonant) Loops

Self resonant loop antennas are relatively large, governed by the intended wavelength of operation. They are mainly used at frequencies above 3.5 MHz where their size is feasible. They can be viewed as a folded dipole split into an open shape, just as a folded dipole is a full-sized loop, bent at two ends and squashed into a line. The loop's shape can be a circle, triangle, square, rectangle, or in fact any closed polygon; the only strict requirement is that its perimeter must be (slightly over) one full-wavelength.

The maximum radiation for a 1 wavelength loop is at right angles to the plane of the loop . At the lower shortwave frequencies a full loop is physically quite large, and for practical reasons must be installed "lying flat", that is the plane of the loop horizontal to the ground, its wires supported at the same height by masts at its several corners. The main beam is upwards. Above 10 MHz, the loop is more frequently "standing up", that is with the plane of the loop vertical, in order to direct its main beam towards the horizon. If feasible, a vertical loop may be rotatable, in order to control the direction of the strongest signal. Compared to a dipole or folded dipole, it transmits slightly less toward the sky or ground, giving it about 1.5 dB higher gain in the two favored horizontal directions.

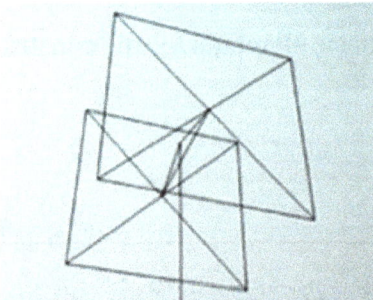

A quad antenna is a resonant loop in a square shape; this one also includes a parasitic element.

Additional gain (and a uni-directional radiation pattern) is usually obtained with an array of such elements either as a driven endfire array or in a Yagi configuration (with all but one loop being parasitic elements). The latter is widely used in amateur radio where it is referred to as a quad antenna.

Loop antennas may be in the shape of a circle, a square or any other closed geometric shape that allows the total perimeter to be one wavelength. The most popular shape in amateur radio is the quad antenna or "quad" (quadrilateral) consists of a resonant loop in a square shape so that it can be constructed of wire strung across a supporting 'X' frame. Usually there are other, additional loops stacked parallel to the first as parasitic elements, that make the composite antenna directional. Other "quads" rotate this 45 degrees to a diamond shape supported on a '+' frame. Triangular loops have also been used. A rectangle twice as high as its width gives a bit more gain than the square loop and also matches 50 ohms directly if used without a reflector.

The polarization of such an antenna is not obvious by looking at the loop itself, but depends on the feed point (where the transmission line is connected), and whether it is being operated as a 1, 2, or 3 wavelength loop. If a vertically oriented loop is fed at the bottom at its 1 wavelength frequency, it will be horizontally polarized; feeding it from the side will make it vertically polarized.

In all of the large loops described above, the antenna's operating frequency is assumed to be at its first resonance, whose corresponding wavelength almost matches the circumference of the loop. Wire size and type of insulation will cause minor shifts in the resonant frequency. Other loop antenna resonances exist near frequencies where the perimeter is 2 wavelengths, 3 wavelengths, etc., with very different radiation patterns.

Low frequency one wavelength loops are sometimes used on higher frequencies where the circumference will be several whole wavelengths. There may be some resonances which may not fall on legally usable frequencies; in cases where the higher resonant frequencies can be used, the feedpoint impedance will also be very different, so operation will require use of an antenna tuner, preferably with a low loss transmission line. Radiation patterns at higher resonances are very different: Most noticeably the maximum radiation is in the plane of the loop, like a small loop, instead of being perpendicular to it.

Small Loops

Small loops are "small" in comparison to their operating wavelength, typically between 5% and 30% of a wavelength in circumference, with transmitting loops tending to be closer to 30%. As with all antennas, smaller antennas are less efficient radiators than larger antennas. However, small loops become practical at lower frequencies where wavelengths are tens to hundreds of meters long, or greater, and full-size loops (the most efficient) and half-wave straight-wire antennas (next-most efficient) become infeasibly large.

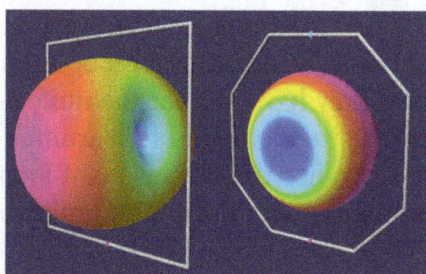

The full wave loop (left) has maximum signal broadside to the wires with nulls off the sides, the small loop (right) has maximum signal in the plane of its wires with nulls broadside to the wires.

A common distinguishing feature of small loops is that their direction of maximum transmission or reception is within the plane of the loop – the opposite of large loops, whose maximum is perpendicular to the plane. In the direction that large loops produce their strongest signals in both transmit and receive, small loops have a null in their pattern.

Carefully designed and built small loops have advantages for reception on frequencies below 10 MHz. Although a small loop's losses can be high, the receiving signal-to-noise ratio may not suffer if the loop's diameter is at least 1 or 2 meters, regardless of frequency. The very high Q rejects off-frequency interference and overload but also dictates that the loop must be carefully tuned to the exact operating frequency. The ability to rotate may help reject either local noise or distant interference, by orienting the "deaf" side of the loop towards the unwanted interference.

Small Transmitting Loops

Size, Shape, Efficiency and Pattern

Small transmitting loops are "small" in comparison to a full-wave loop, but considerably larger than the small receiving loop, and unlike receiving loops must be "scaled-up" for longer wavelengths.

They are typically used on frequencies between 3–30 MHz. They usually consist of a single turn of large diameter conductor, and are typically round or octagonal to provide maximum enclosed area for a given perimeter. The smaller of these loops are much less efficient than full-sized self-resonant loops, but where space is at a premium the smaller loops can provide effective communications. Loop antennas are relatively easy to build.

A loop antenna for amateur radio under construction.

A small transmitting loop antenna, also known as a magnetic loop, with a circumference 10% of a wavelength or less, will have a relatively constant current distribution along the conductor, and the main lobe will be in the plane of the loop. Loops of any size between 10% and 100% of a wavelength in circumference can be built and tuned to resonance with series reactance. A capacitor is required for a circumference less than a half wave, an inductor for loops more than a half wave and less than a full wave. Loops in this size range may have neither the uniform current of the small loop, nor the double peaked current of the full sized loop and thus cannot be analyzed using the concepts developed for the small receiving loops nor the self resonant loop antennas. Performance is best determined with NEC analysis. Antennas within this size range include the halo and the GoCWT (Edginton) loop.

All small transmitting loops work even better for receiving.

Matching to the Transmitter

In addition to other common impedance matching techniques such as a gamma match, transmitting loops are sometimes impedance matched by connecting the feedline to a smaller *feed loop* inside the area surrounded by the main loop. Typical feed loops are 1/8 to 1/5 the size of the antenna's main loop. The combination is in effect a transformer, with power in the near-field inductively coupled from the feed loop to the main loop, which itself is connected to the resonating capacitor and is responsible for radiating most of the power.

Use for Land-mobile Radio

Small loops are used in land-mobile radio (mostly military) at frequencies between 3–7 MHz, because of their ability to direct energy upwards, unlike a conventional whip antenna. This enables Near Vertical Incidence Skywave (NVIS) communication up to 300 km in mountainous regions. In this case a typical radiation efficiency of around 1% is acceptable because signal paths can be established with 1 Watt of radiated power or less when a transmitter generating 100 Watts is used. In military use, the antenna may be built using a one or two conductors 1–2 inches in diameter. The loop itself is typically 6 feet in diameter.

Power Limits

One practical issue with small loops as transmitting antennas is that the loop not only has a very large current going through it, but also has a very high voltage across the capacitor, typically kilo-Volts when fed with only a few watts of transmitter power. This requires a rather expensive and physically large resonating capacitor with a large breakdown voltage, in addition to having minimal dielectric loss (normally requiring an air-gap capacitor). In addition to making the geometric loop larger, efficiency may be increased by using larger conductors or other measures to reduce the conductor's loss resistance. However, lower loss means higher Q and even greater voltage on the capacitor.

This problem is more serious than with a vertical or dipole antenna that is short compared to a wavelength. There matching using a loading coil also generates a high voltage at the antenna ends. However, unlike with capacitors, the voltage across a physically large inductor is generally not an issue.

Small Receiving Loops

Small loop antenna used for receiving, consisting of about 10 turns around a 12 cm × 10 cm rectangle.

Although a full 2.7 meters in diameter, this receiving antenna is a "small" loop compared to LF and MF wavelengths.

If the dimensions of a loop antenna are much less than the intended wavelengths, then the antenna is a small loop antenna. For a given loop area, the length of the conductor (and thus its net loss resistance) is minimized if the shape is a circle, making a circle the optimal shape for small loops. Small receiving loops are typically used below 3 MHz where human-made and natural atmospheric noise dominate. Thus the signal-to-noise ratio of the received signal will not be adversely affected by low efficiency as long as the loop is not excessively small.

A typical diameter of receiving loops with "air centers" is between 30 cm and 1 meter. To increase the magnetic field in the loop and thus its efficiency, while greatly reducing size, the coil of wire is often wound around a ferrite rod magnetic core; this is called a *ferrite loop* antenna. Such ferrite loop antennas are used in almost all AM broadcast receivers with the exception of car radios; the antenna is then usually contained inside the radio's chassis. These antennas are also used for radio direction finding.

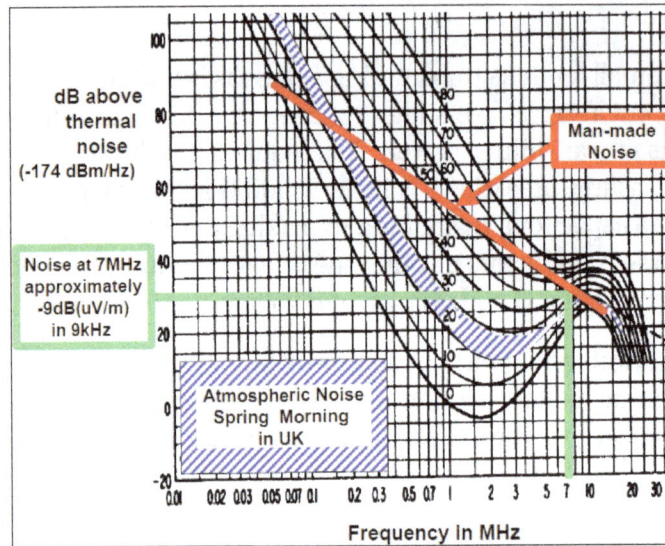

Amount of atmospheric noise for LF, MF, and HF spectrum according CCIR 322.

The radiation resistance R_R of a small loop is generally much smaller than the loss resistance R_L due to the conductors comprising the loop, leading to a poor antenna efficiency. Consequently, most of the power delivered to a small loop antenna will heat in the loss resistance rather than do useful work.

So much wasted power is not acceptable in a transmitting antenna, however in a receiving antenna the inefficiency is not important at frequencies below about 15 MHz. At these lower frequencies, atmospheric noise (static) and man-made noise (radio frequency interference) even in the weak signal from an inefficient antenna are far above the internal *thermal* or *Johnson noise* present in the radio receiver's circuits, so the weak signal from a loop antenna can be amplified without degrading the signal-to-noise ratio.

For example, at 1 MHz the man-made noise might be 55 dB above the thermal noise floor. If a small loop antenna's loss is 50 dB (as if the antenna included a 50 dB attenuator) the electrical inefficiency of that antenna will have little influence on the receiving system's signal-to-noise ratio.

In contrast, at quieter frequencies at about 20 MHz and above, an antenna with a 50 dB loss could degrade the received signal-to-noise ratio by up to 50 dB, resulting in terrible performance.

Magnetic vs. Electrical Antennas

The small loop antenna is known as a *magnetic loop* since it behaves electrically as a coil (inductor). It couples to the magnetic field of the radio wave in the region near the antenna, in contrast to monopole and dipole antennas which couple to the electric field of the wave. In a receiving antenna (the main application of small loops) the oscillating magnetic field of the incoming radio wave induces a current in the wire winding by Faraday's law of induction.

Radiation Pattern and Polarization

Surprisingly, the radiation and receiving pattern of a small loop is quite opposite that of a large self resonant loop (whose circumference is close to one wavelength). Since the loop is much smaller than a wavelength, the current at any one moment is nearly constant round the circumference. By symmetry it can be seen that the voltages induced in the loop windings along the plane of the loop, will cancel each other when a perpendicular signal arrives on the loop axis. Therefore, there is a *null* in that direction. Instead, the radiation pattern peaks in directions lying in the plane of the loop, because signals received from sources in that plane do not quite cancel owing to the phase difference between the arrival of the wave at the near side and far side of the loop. Increasing that phase difference by increasing the size of the loop has a large impact in increasing the radiation resistance and the resulting antenna efficiency.

Another way of looking at a small loop as an antenna is to consider it simply as an inductive coil coupling to the magnetic field in the direction *perpendicular* to plane of the coil, according to Ampère's law. Then consider a propagating radio wave also perpendicular to that plane. Since the magnetic (and electric) fields of an electromagnetic wave in free space are transverse (no component in the direction of propagation), it can be seen that this magnetic field and that of a small loop antenna will be at right angles, and thus not coupled. For the same reason, an electromagnetic wave propagating within the plane of the loop, with its magnetic field perpendicular to that plane, *is* coupled to the magnetic field of the coil. Since the transverse magnetic and electric fields of a propagating electromagnetic wave are at right angles, the electric field of such a wave is also in the plane of the loop, and thus the antenna's *polarization* (which is always specified as being the orientation of the electric, not the magnetic field) is said to be in that plane.

Thus mounting the loop in a horizontal plane will produce an omnidirectional antenna which is horizontally polarized; mounting the loop vertically yields a weakly directional antenna with vertical polarization and sharp nulls along the axis of the loop.

Receiver Input Tuning

Since a small loop antenna is essentially a coil, its electrical impedance is inductive, with an inductive reactance much greater than its radiation resistance. In order to couple to a transmitter or receiver, the inductive reactance is normally canceled with a parallel capacitance. Since a good loop antenna will have a high Q factor, this capacitor must be variable and is adjusted along with the receiver's tuning.

Small loop receiving antennas are also almost always resonated using a parallel plate capacitor, which makes their reception narrow-band, sensitive only to a very specific frequency. This allows the antenna, in conjunction with a (variable) tuning capacitor, to act as a tuned input stage to the receiver's front-end, in lieu of a preselector.

Insensitivity to Locally Generated Interference

Due to its direct coupling to the magnetic field, unlike most other antenna types, the small loop is relatively insensitive to electric-field noise from nearby sources. No matter how close the electrical interference is to the loop, its effect will not be much greater than if it were a quarter wavelength away. This is valuable since most sources of interference with radio frequency content, such as sparking at commutators or corona discharge, directly produce electric fields in the near-field (much less than a wavelength from the source). Since it is in the AM broadcast band and lower frequencies generally where these small loops are used, the near field region is physically quite large (on the order of 30 m, or 100 feet). This provides a considerable advantage for using an antenna which is relatively insensitive to the main interference sources encountered in that frequency range.

The same principle makes a small loop particularly sensitive to sources of *magnetic* noise in its near field. Likewise, a Hertzian (short) dipole couples directly with the electric field and is relatively immune to locally produced magnetic noise. However at radio frequencies nearby sources of magnetic interference are generally not an issue. In either case the small antenna's immunity does not extend to noise sources outside of the near field: Noise sources over one wavelength distant, whether originating as an electric or magnetic field, are received simply as electromagnetic waves. Noise from outside any antenna's near field will be received equally well by any antenna sensitive to a radio transmitter from the direction of that noise source.

Direction Finding with Small Loops

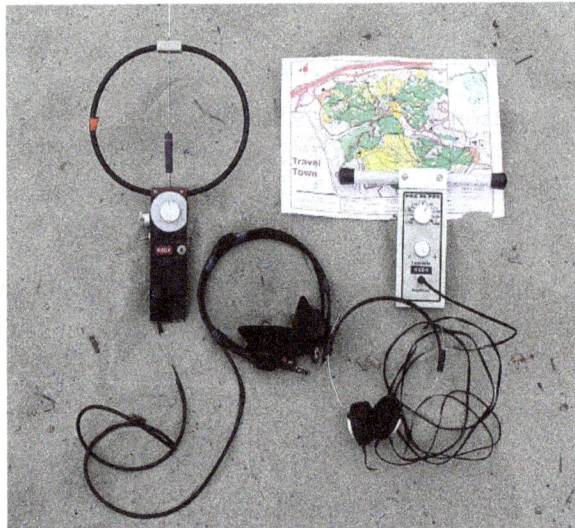

Loop antenna, receiver, and accessories used in amateur radio
direction finding at 80 meter wavelength (3.5 MHz).

Since the directional response of small loop antennas includes a sharp null in the direction normal to the plane of the loop, they are used in radio direction finding at longer wavelengths.

The procedure is to rotate the loop antenna to find the direction where the signal vanishes – the "null" direction. Since the null occurs at two opposite directions along the axis of the loop, other means must be employed to determine which side of the antenna the "nulled" signal is on. One method is to rely on a second loop antenna located at a second location, or to move the receiver to that other location, thus relying on triangulation.

Instead of triangulation, a second dipole or vertical antenna can be electrically combined with a loop or a loopstick antenna. Called a *sense antenna*, connecting and matching the second antenna changes the combined radiation pattern to a cardioid, with a null in only one (less precise) direction. The general direction of the transmitter can be determined using the sense antenna, and then disconnecting the sense antenna returns the sharp nulls in the loop antenna pattern, allowing a precise bearing to be determined.

AM Broadcast Receiving Antennas

Small loop antennas are lossy and inefficient for transmitting, but they can be practical receiving antennas for frequencies below 10 MHz. Especially in the mediumwave (520–1610 kHz) band and below, where wavelength-sized antennas are infeasibly large, and the antenna inefficiency is irrelevant, due to large amounts of atmospheric noise.

AM broadcast receivers (and other low frequency radios for the consumer market) typically use small loop antennas, even when a telescoping antenna may be attached for FM reception. A variable capacitor connected across the loop forms a resonant circuit that also tunes the receiver's input stage as that capacitor tracks the main tuning. A multiband receiver may contain tap points along the loop winding in order to tune the loop antenna at widely different frequencies.

In AM radios built prior to the discovery of ferrite in the mid-20th century, the antenna might consist of dozens of turns of wire mounted on the back wall of the radio – a *planar helical antenna* – or a separate, rotatable, furniture-sized rack looped with wire – a *frame antenna*.

Ferrite

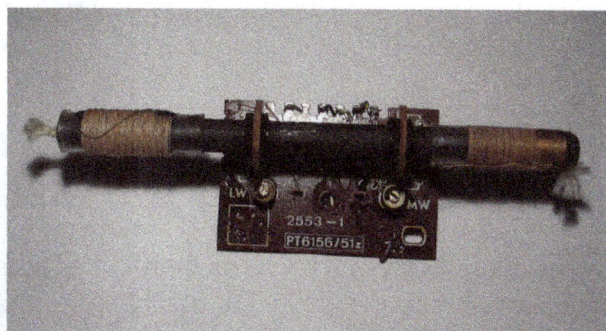

Ferrite loopstick antenna from an AM radio having two windings, one for long wave and one for medium wave (AM broadcast) reception. About 10 cm long. Ferrite antennas are usually enclosed inside the radio receiver.

Ferrite loop antennas are made by winding fine wire around a ferrite rod. They are almost universally used in AM broadcast receivers. Other names for this type of antenna are loopstick, ferrite rod antenna or aerial, ferroceptor, or ferrod antenna. Often at shortwave frequencies Litz wire is used for the winding to reduce skin effect losses. Elaborate "basket weave" patterns are used at all frequencies to reduce self-capacitance in the coil and raise the loop self-resonance above the operating frequency, consequently improving the loop Q factor.

The ferrite increases the magnetic permeability and acts as a low loss magnetic conductor – much better than air. This greater conductance channels thousands of times more magnetic power through the rod, and hence through the loop, allowing the physically small antenna to have a larger effective area.

Loop-like Antennas

Some antennas look very much like loops, but are either not continuous loops, or are designed to couple with the inductive near-field – over distances of a meter or two – rather than to transmit or receive long-distance electromagnetic waves in the radiative far-field.

Halo Antennas

Although it has a superficially similar appearance, the so-called halo antenna is not technically a loop since it possesses a break in the conductor opposite the feed point. It is better analyzed as a dipole which has been bent into a circle. However, if we consider that small currents flow between the closely spaced ends of the dipole, the halo can be viewed as a small transmitting loop in the limiting case where the resonating capacitor has been reduced to a very small value as the circumference has increased to about one half wave.

RFID Coils

Strictly speaking, RFID tags and readers interact by induction rather than transmission waves, and so are not antennas.

These systems do operate at radio frequencies, and do involve the use of small loops which are called "antennas" in the trade. Although these small loops are sometimes indistinguishable from the small loop antennas, such systems are not designed to transmit or receive signal waves (electromagnetic waves), and can only operate over short distances. They are near field systems involving alternating magnetic fields, and may be analyzed as poorly coupled transformer windings; their performance criteria are dissimilar to radio antennas.

SHORT DIPOLE

A short dipole is a simple wire antenna. One end of it is open-circuited and the other end is fed with AC source. This dipole got its name because of its length.

Frequency Range

The range of frequency in which short dipole operates is around 3KHz to 30MHz. This is mostly used in low frequency receivers.

Construction and Working of Short Dipole

The Short dipole is the dipole antenna having the length of its wire shorter than the wavelength. A voltage source is connected at one end while a dipole shape is made, i.e., the lines are terminated at the other end.

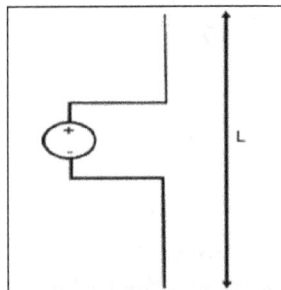

The circuit diagram of a short dipole with length L is shown. The actual size of the antenna does not matter. The wire that leads to the antenna must be less than one-tenth of the wavelength. That is,

$$L < \frac{\lambda}{10}$$

Where

- L is the length of the wire of the short dipole.

- λ is the wavelength.

Another type of short dipole is infinitesimal dipole, whose length is far less than its wave length. Its constructiion is similar to it, but uses a capacitor plate.

Infinitesimal Dipole

A dipole whose length is far less than wavelength is infitesimal dipole. This antenna is actually impractical. Here, the length of the dipole is less than even fiftith part of the wavelength.

The length of the dipole, $\Delta l << \lambda$. Where, λ is the wavelength.

$$\Delta l = \frac{\lambda}{50}$$

Hence, this is the infinitely small dipole, as the name implies.

As the length of these dipoles is very small, the current flow in the wire will be dI. These wires are generally used with capacitor plates on both sides, where low mutual coupling is needed. Because of the capacitor plates, we can say that uniform distribution of current is present. Hence the current is not zero here.

The capacitor plates can be simply conductors or the wire equivalents. The fields radiated by the radial currents tend to cancel each other in the far field so that the far fields of the capacitor plate antenna can be approximated by the infinitesimal dipole.

Radiation Pattern

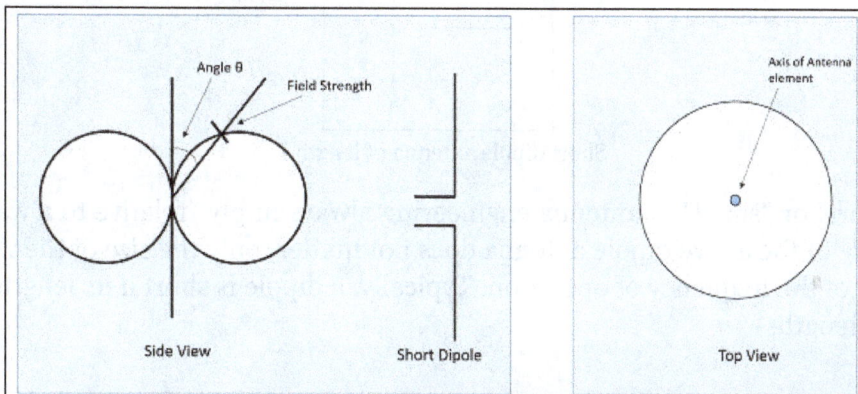

Short dipole antenna radiation pattern.

The radiation pattern of a short dipole and infinitesimal dipole is similar to a half wave dipole. If the dipole is vertical, the pattern will be circular. The radiation pattern is in the shape of "figure of eight" pattern, when viewed in two-dimensional pattern.

Advantages

The following are the advantages of short dipole antenna:

- Ease of construction, due to small size.
- Power dissipation efficiency is higher.

Disadvantages

The following are the disadvantages of short dipole antenna:

- High resistive losses.
- High power dissipation.
- Low Signal-to-noise ratio.
- Radiation is low.
- Not so efficient.

Applications

The following are the applications of short dipole antenna:

- Used in narrow band applications.
- Used as an antenna for tuner circuits.

The short dipole antenna is the simplest of all antennas. It is simply an open-circuited wire, fed at its center as shown in figure.

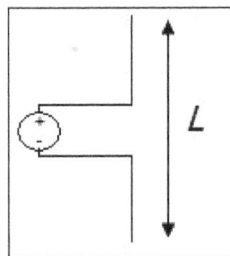

Short dipole antenna of length L.

The words "short" or "small" in antenna engineering always imply "relative to a wavelength". So the absolute size of the above dipole antenna does not matter, only the size of the wire relative to the wavelength of the frequency of operation. Typically, a dipole is short if its length is less than a tenth of a wavelength:

$$L < \frac{\lambda}{10}$$

If the short dipole antenna is oriented along the z-axis with the center of the dipole at z=0, then the current distribution on a thin, short dipole is given by:

$$I(z) = I_0 \left(1 - \frac{2|z|}{L} \right)$$

The current distribution is plotted in figure. Note that this is the amplitude of the current distribution; it is oscillating in time sinusoidally at frequency f.

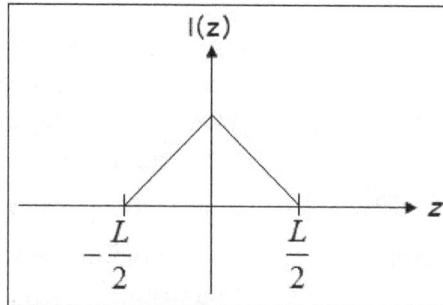

Current distribution along a short dipole antenna.

The fields radiated from the short dipole antenna in the far field are given by:

$$E_\theta = \frac{j\eta k I_0 L e^{-jkr}}{8\pi r} \sin\theta$$

$$H_\phi = \frac{E_\theta}{\eta}$$

$$E_r = H_r = E_\phi = H_\theta = 0$$

The above equations can be broken down and understood somewhat intuitively. First, note that in the far-field, only the E_θ and H_ϕ fields are nonzero. Further, these fields are orthogonal and in-phase. Further, the fields are perpendicular to the direction of propagation, which is always in the $\hat{r}$ direction (away from the antenna). Also, the ratio of the E-field to the H-field is given by η (the intrinsic impedance of free space).

This indicates that in the far-field region the fields are propagating like a plane-wave.

Second, the fields die off as 1/r, which indicates the power falls of as:

$$P(r) \propto \frac{1}{r^2}$$

Third, the fields are proportional to L, indicated a longer dipole will radiate more power. This is true as long as increasing the length does not cause the short dipole assumption to become invalid. Also, the fields are proportional to the current amplitude I_0, which should make sense.

The exponential term:

$$e^{-jkr}$$

describes the phase-variation of the wave versus distance. The parameter k is known as the wavenumber. Note also that the fields are oscillating in time at a frequency f in addition to the above spatial variation.

Finally, the spatial variation of the fields as a function of direction from the antenna are given by $\sin\theta$. For a vertical antenna oriented along the z-axis, the radiation will be maximum in the x-y plane. Theoretically, there is no radiation along the z-axis far from the antenna.

Directivity, Impedance and other Properties of the Short Dipole Antenna

The directivity of the center-fed short dipole antenna depends only on the $\sin\theta$ component of the fields. It can be calculated to be 1.5 (1.76 dB), which is very low for realizable (physical or non-theoretical) antennas. Since the fields of the short dipole antenna are only a function of the polar angle, they have no azimuthal variation and hence this antenna is characterized as omnidirectional. The Half-Power Beamwidth is 90 degrees.

The polarization of this antenna is linear. When evaluated in the x-y plane, this antenna would be described as vertically polarized, because the E-field would be vertically oriented (along the z-axis).

We now turn to the input impedance of the short dipole, which depends on the radius a of the dipole. Recall that the impedance Z is made up of three components, the radiation resistance, the loss resistance, and the reactive (imaginary) component which represents stored energy in the fields:

$$Z = R_{rad} + R_{loss} + jX$$

The radiation resistance can be calculated to be:

$$R_{rad} = 20\pi^2 \left(\frac{L}{\lambda}\right)^2$$

The resistance representing loss due to the finite-conductivity of the antenna is given by:

$$R_{loss} = \frac{L}{6\pi a}\sqrt{\frac{\pi f u}{2\sigma}}$$

In the above equation σ represents the conductivity of the dipole (usually very high, if made of metal). The frequency f come into the above equation because of the skin effect. The reactance or imaginary part of the impedance of a dipole is roughly equal to:

$$X = \frac{-120\lambda}{\pi L}\left(\ln\left(\frac{L}{2a}\right)-1\right)$$

As an example, assume that the radius is 0.001 λ and the length is 0.05 λ. Suppose further that this antenna is to operate at f=3 MHz, and that the metal is copper, so that the conductivity is 59,600,000 S/m.

The radiation resistance is calculated to be 0.49 Ohms. The loss resistance is found to be 4.83 mOhms (milli-Ohms), which is approximatley negligible when compared to the radiation resistance. However, the reactance is 1695 Ohms, so that the input resistance is Z=0.49 + j1695. Hence, this antenna would be very difficult to have proper impedance matching. Even if the reactance could be properly cancelled out, very little power would be delivered from a 50 Ohm source to a 0.49 Ohm load.

For short dipole antennas that are smaller fractions of a wavelength, the radiation resistance becomes smaller than the loss resistance, and consequently this antenna can be very inefficient.

The bandwidth for short dipoles is difficult to define. The input impedance varies wildly with frequency because of the reactance component of the input impedance. Hence, these antennas are typically used in narrowband applications.

MONOPOLE

A monopole antenna is a class of radio antenna consisting of a straight rod-shaped conductor, often mounted perpendicularly over some type of conductive surface, called a ground plane. The driving signal from the transmitter is applied, or for receiving antennas the output signal to the receiver is taken, between the lower end of the monopole and the ground plane. One side of the antenna feedline is attached to the lower end of the monopole, and the other side is attached to the ground plane, which is often the Earth. This contrasts with a dipole antenna which consists of two identical rod conductors, with the signal from the transmitter applied between the two halves of the antenna.

A typical mast radiator monopole antenna of an AM radio station.

The monopole is a resonant antenna; the rod functions as an open resonator for radio waves, oscillating with standing waves of voltage and current along its length. Therefore, the length of the antenna is determined by the wavelength of the radio waves it is used with. The most common form is the *quarter-wave monopole*, in which the antenna is approximately one quarter of the wavelength of the radio waves. The monopole antenna was invented in 1895 by radio pioneer Guglielmo Marconi; for this reason it is sometimes called the *Marconi antenna*. Common types of monopole antenna are the whip, rubber ducky, helical, random wire, umbrella, inverted-L and T-antenna, inverted-F, mast radiator, and ground plane antennas.

The load impedance of the quarter-wave monopole is half that of the dipole antenna or 37.5+j21.25 ohms.

The mast itself is connected to the transmitter and radiates the radio waves. It is mounted on a ceramic insulator to isolate it from the ground. The other terminal of the transmitter is connected to a ground system consisting of cables buried under the field.

Radiation Pattern

Like a dipole antenna, a monopole has an omnidirectional radiation pattern: it radiates with equal power in all azimuthal directions perpendicular to the antenna. However, the radiated power varies with elevation angle, with the radiation dropping off to zero at the zenith of the antenna axis. It radiates vertically polarized radio waves.

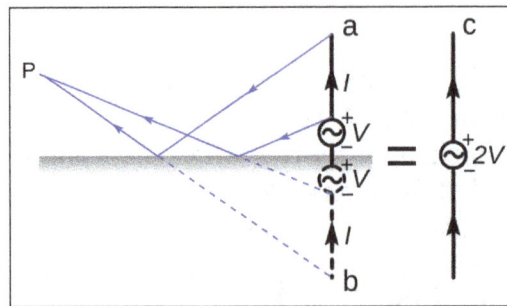

Showing the monopole antenna has the same radiation pattern over perfect ground as a dipole in free space with twice the voltage

A monopole can be visualized (right) as being formed by replacing the bottom half of a vertical dipole antenna (c) with a conducting plane (ground plane) at right-angles to the remaining half. If the ground plane is large enough, the radio waves from the remaining upper half of the dipole (a) reflected from the ground plane will seem to come from an image antenna (b) forming the missing half of the dipole, which adds to the direct radiation to form a dipole radiation pattern. So the pattern of a monopole with a perfectly conducting, infinite ground plane is identical to the top half of a dipole pattern, with its maximum radiation in the horizontal direction, perpendicular to the antenna. Because it radiates only into the space above the ground plane, or half the space of a dipole antenna, a monopole antenna will have a gain of twice (3 dB greater than) the gain of a similar dipole antenna, and a radiation resistance half that of a dipole. Since a half-wave dipole has a gain of 2.19 dBi and a radiation resistance of 73 ohms, a quarter-wave monopole, the most common type, will have a gain of 2.19 + 3 = 5.19 dBi and a radiation resistance of about 36.8 ohms if it is mounted above a good ground plane.

The general effect of electrically small ground planes, as well as imperfectly conducting earth grounds, is to tilt the direction of maximum radiation up to higher elevation angles.

Types

The ground plane used with a monopole may be the actual earth; in this case the antenna is mounted on the ground and one side of the feedline is connected to an earth ground at the base of the antenna. This design is used for the mast radiator antennas employed in radio broadcasting at low frequencies, as well as other low frequency antennas such as the T-antenna and umbrella antenna. At VHF and UHF frequencies the size of the ground plane needed is smaller, so artificial ground planes are used to allow the antenna to be mounted above the ground. A common type of monopole antenna at these frequencies consists of a quarter-wave whip antenna with a ground plane consisting of several wires or rods radiating horizontally or diagonally from its base; this is called a ground-plane antenna. At gigahertz frequencies the metal surface of a car roof or airplane body makes a good ground plane, so car cell phone antennas consist of short whips mounted on the roof, and aircraft communication antennas frequently consist of a short conductor in an aerodynamic fairing projecting from the fuselage; this is called a *blade antenna*. The most common antenna used in mobile phones is the inverted-F antenna, which is a variant of the inverted-L monopole. Bending over the antenna saves space and keeps the it within the bounds of the mobile's case but the antenna then has a very low impedance. To improve the match the antenna is not fed from the end, rather some intermediate point, and the end is grounded instead. The quarter-wave whip and rubber ducky antennas used with handheld radios such as walkie-talkies and cell phones are also monopole antennas. These don't use a ground plane, and the ground side of the transmitter is just connected to the ground connection on its circuit board. The hand and body of the person holding them may function as a rudimentary ground plane.

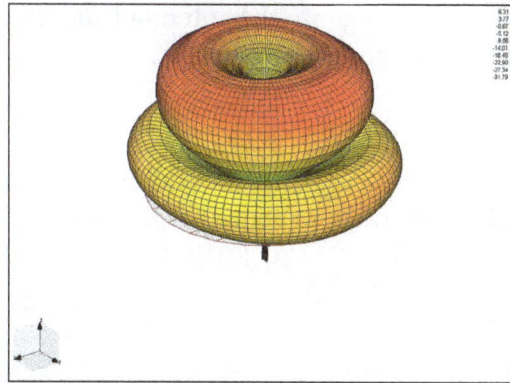

Radiation pattern of 3/2 wavelength monopole. Monopole antennas up to 1/4 wavelength long have a single "lobe", with field strength declining monotonically from a maximum in the horizontal direction, but longer monopoles have more complicated patterns with several conical "lobes" (radiation maxima) directed at angles into the sky.

Sometimes, monopole antennas are printed on a dielectric substrate to make it less fragile and they may be fabricated easily using the printed circuit board technologies. Such antennas are known as printed monopole antennas. They are suitable for various applications such as RFID, WLAN.

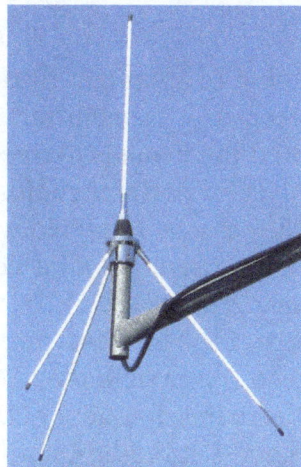

VHF ground plane antenna, a type of monopole antenna used at high frequencies. The three conductors projecting downward are the ground plane.

Monopole Broadcasting Antennas

When used for radio broadcasting, the radio frequency power from the broadcasting transmitter is fed across the base insulator between the tower and a ground system. The ideal ground system for AM broadcasters comprises at least 120 buried copper or phosphor bronze radial wires at least one-quarter wavelength long and a ground-screen in the immediate vicinity of the tower. All the ground system components are bonded together, usually by welding, brazing or using coin silver solder to help reduce corrosion. Monopole antennas that use guy-wires for support are called masts in some countries. In the United States, the term "mast" is generally used to describe a pipe supporting a smaller antenna, so both self-supporting and guy-wire supported radio antennas are simply called monopoles if they stand alone. If multiple monopole antennas are used in order to control the direction of radio frequency (RF) propagation, they are called directional antenna arrays.

In the United States, the Federal Communications Commission (FCC) requires that the transmitter power input to the antenna be measured and maintained. The power input is calculated as the square of the measured current, i, flowing into the antenna from the transmission line multiplied by the real part of the antenna's feed-point impedance, r.

$$P = i^2 r$$

This impedance is periodically measured to verify the stability of the antenna and ground system. Normally, an impedance matching network matches the impedance of the antenna to the impedance of the transmission line feeding it.

Examples of monopole antennas are:

- The whip antenna.

- The mast radiator (when isolated from the ground and bottom-fed).

Monopole antennas have become one of the components of mobile and Internet networks across the globe. Their relative low cost and fast installation makes them an obvious choice for developing countries.

Whip Antenna

A whip antenna is an antenna consisting of a straight flexible wire or rod. The bottom end of the whip is connected to the radio receiver or transmitter. The antenna is designed to be flexible so that it does not break easily, and the name is derived from the whip-like motion that it exhibits when disturbed. Whip antennas for portable radios are often made of a series of interlocking telescoping metal tubes, so they can be retracted when not in use. Longer ones, made for mounting on vehicles and structures, are made of a flexible fiberglass rod around a wire core and can be up to 35 ft (10 m) long. The length of the whip antenna is determined by the wavelength of the radio waves it is used with. The most common type is the *quarter-wave whip*, which is approximately one-quarter of a wavelength long. Whips are the most common type of monopole antenna, and are used in the higher frequency HF, VHF and UHF radio bands. They are widely used as the antennas for hand-held radios, cordless phones, walkie-talkies, FM radios, boom boxes, and Wi-Fi enabled devices, and are attached to vehicles as the antennas for car radios and two-way radios for wheeled vehicles and for aircraft. Larger versions mounted on roofs and radio masts are used as base station antennas for police, fire, ambulance, taxi, and other vehicle dispatchers.

Radiation Pattern

The whip antenna is a monopole antenna, and like a vertical dipole has an omnidirectional radiation pattern, radiating equal radio power in all azimuthal directions (perpendicular to the antenna's axis), with the radiated power falling off with elevation angle to zero on the antenna's axis. Whip antennas 1/4 wavelength long or less (the most common type) have a single main lobe, with field strength maximum in horizontal directions, falling monotonically to zero on the axis. Antennas longer than a quarter wavelength have patterns consisting of several conical "lobes"; with radiation maxima at several elevation angles; the longer the electrical length of the antenna, the more lobes the pattern has.

Vertical whip antennas are widely used for nondirectional radio communication on the surface of the Earth, where the direction to the transmitter (or the receiver) is unknown or constantly changing, for example in portable FM radio receivers, walkie-talkies, and two-way radios in vehicles. This is because they transmit (or receive) equally well in all horizontal directions, while radiating little radio energy up into the sky where it is wasted.

Three large fiberglass whips mounted on a mast.

Length

Whip antennas are normally designed as resonant antennas; the rod acts as a resonator for radio waves, with standing waves of voltage and current reflected back and forth from its ends. Therefore, the length of the antenna rod is determined by the wavelength of the radio waves used. The most common length is approximately one-quarter of the wavelength, called a "quarter-wave whip". For example, the common quarter-wave whip antennas used on FM radios in the USA are approximately 75 cm long, which is roughly one-quarter the length of radio waves in the FM radio band, which are 2.78 to 3.41 meters long. Half-wave antennas are also common.

Gain and Radiation Resistance

A quarter wave vertical antenna working against a perfect infinite ground will have a gain of 5.19 dBi and about 36.8 ohms of radiation resistance. Whips mounted on vehicles use the metal skin of the vehicle as a ground plane. In hand-held devices usually no explicit ground plane is provided, and the ground side of the antenna's feed line is just connected to the ground on the device's circuit board. Therefore, the radio itself, and possibly the user's hand, serves as a rudimentary ground plane. Since these are no larger than the size of the antenna itself, the combination of whip and radio often functions more as an asymmetrical dipole antenna than as a monopole antenna. The gain will suffer somewhat compared to a half wave metallic dipole or a whip with a well defined ground plane.

Ground Plane Antenna

With stationary whips mounted on structures, an artificial "ground plane" consisting of three or four rods a quarter-wavelength long extending horizontally from the base of the whip is often used. This provides a stable input impedance and pattern by helping prevent RF currents in the supporting mast and along the outside of the feed line. This type of antenna is called a *ground plane antenna*. Often the ground plane rods are sloped downward toward the ground, which lowers the main lobe of the radiation pattern and increases the normal 36.8 ohm radiation resistance closer to 50 ohms to provide a better impedance match with standard 50 ohm coaxial cable feedline.

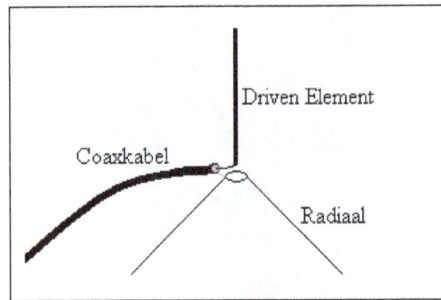

GP Antenna.

Electrically Short Whips

A rubber ducky antenna, a common type of electrically short whip, on a handheld UHF CB transceiver. With rubber sheath (left) removed.

To reduce the length of a whip antenna to make it less cumbersome, an inductor (loading coil) is often added in series with it. This allows the antenna to be made much shorter than the normal length of a quarter-wavelength, and still be resonant, by cancelling out the capacitive reactance of the short antenna. The coil is added at the base of the whip (called a base-loaded whip) or occasionally in the middle (center-loaded whip). In the most widely used form, the rubber ducky antenna, the loading coil is integrated with the antenna itself by making the whip out of a narrow helix of springy wire. The helix distributes the inductance along the antenna's length, improving the radiation pattern, and also makes it more flexible. Another alternative occasionally used to shorten the antenna is to add a "capacity hat", a metal screen or radiating wires, at the end. However all these electrically short whips have lower gain than a full length quarter-wave whip.

Cellphone whip antenna with base loading coil on car.

Collection of walkie-talkies with electrically short whips. Units on ends and small one in foreground have "rubber ducky" antennas.

Multi-band operation is possible with coils at about one-half or one-third and two-thirds that do not affect the aerial much at the lowest band, but it creates the effect of stacked dipoles at a higher band (usually ×2 or ×3 frequency).

FM whip antenna on portable receiver.

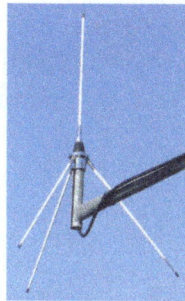

Ground plane antenna.

At higher frequencies (2.4 GHz, but military whips for 50 MHz to 80 MHz band exist, and are standard issue for the SINCGARS radio in the 30–88 MHz range), the feed coax can go up the centre of a tube. The insulated junction of the tube and whip is fed from the coax and the lower tube end where coax cable enters has an insulated mount. This kind of vertical whip is a full dipole and thus needs no ground plane. It generally works better several wavelengths above ground, hence the limitation normally to microwave bands.

Mast Radiator

A mast radiator (or radiating tower) is a radio mast or tower in which the entire structure functions as an antenna. This design, first used in radiotelegraphy stations in the early 1900s, is commonly used for transmitting antennas operating at low frequencies, in the VLF, LF and MF ranges, in particular those used for AM broadcasting. The metal mast is electrically connected to the transmitter. Its base is usually mounted on a nonconductive support to insulate it from the ground. A mast radiator is a form of monopole antenna.

Design Considerations

Design

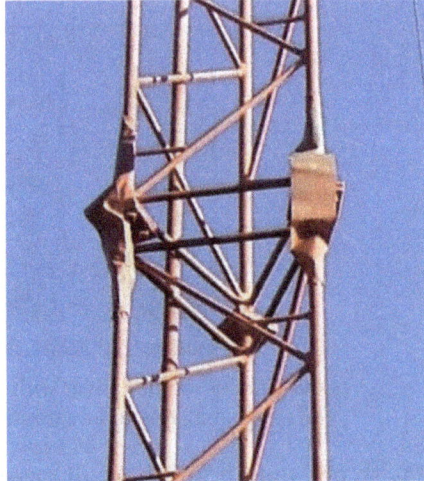

To ensure that the mast acts as a single conductor, the separate structural sections of the mast are connected electrically by copper jumpers or "fusion" (arc) welds across the mating flanges.

Most mast radiators are built as guyed masts insulated from the ground at the base. Steel lattice masts of triangular cross-section are the most common type. Square lattice masts and tubular masts are also sometimes used. To ensure that the tower is a continuous conductor, the tower's structural sections are electrically bonded at the joints by short copper jumpers which are soldered to each side.

Base feed: Radio frequency power is fed to the mast by a wire attached to it, which comes from a matching network inside the "antenna tuning hut" at right. The brown ceramic insulator at the base keeps the mast isolated from the ground. On the left there is an earthing switch and a spark gap for lightning protection.

At its base, the mast is usually mounted on a thick ceramic insulator, which has the compressive strength to support the tower's weight and the dielectric strength to withstand the high RF voltage applied by the transmitter. The RF power to drive the antenna is supplied by a matching network, usually housed in an antenna tuning hut next to the mast, and the cable supplying the current is simply bolted to the tower. The actual transmitter is usually located in a separate building, which supplies RF power to the antenna tuning hut via a transmission line.

The Blosenbergturm in Beromünster, Switzerland - a radiating tower insulated against ground.

Free-standing towers are also used as radiating structures. These towers can have a triangular or a square cross section, with each leg supported on an insulator. One of the best-known radiating towers is the Blosenbergturm in Beromünster, Switzerland. Fiberglass masts are sometimes used for small constructions.

Mast Height

The ideal height of a mast radiator depends on transmission frequency, demographics for the location, and terrain. For radio systems in the longwave and mediumwave range, the value of the height should be in the range between one sixth and five eighths of the wavelength, with preferred values at the quarter or the half of the radiated wavelength. When this is not possible, masts with a loading coil, 'capacity hat' or similar loading arrangement are used.

The height of the mast determines the radiation properties. For high power transmitters in the MW range, masts with heights around half of the radiated wavelength are preferred because they focus the radiated power better to the ground than structures with heights of quarter wavelengths, which are preferred for economical reasons for low power medium wave transmitters. A focus of radiated power towards the ground is much desired on frequencies below 3 megahertz, because ground-wave propagation is very stable. Masts longer than five eighths of the wavelength are normally not used, because they show bad vertical radiation patterns, so masts for mediumwave transmitters do not normally exceed 300 metres. For longwave transmitters, however, the construction of masts with heights of half-wave wavelength is generally not economically viable and in most cases impossible. The only longwave radio mast with a height of the half length of the radiated wavelength was the Warszawa Radio Mast at Konstantynów, Poland. At the time of its collapse in 1991 it was the tallest manmade structure in the world, at 646.38 metres (2,120.67 ft) tall, for a wavelength of 1292.76 metres (frequency 232 kHz). For frequencies below longwave, masts are electrically enlarged by loading coils or capacity hats on the top, because masts of even quarter wavelength would be too high to be practical.

Feed Arrangements

Typical 200 foot (61 m) triangular guyed lattice mast of an AM radio station in Mount Vernon, Washington, US.

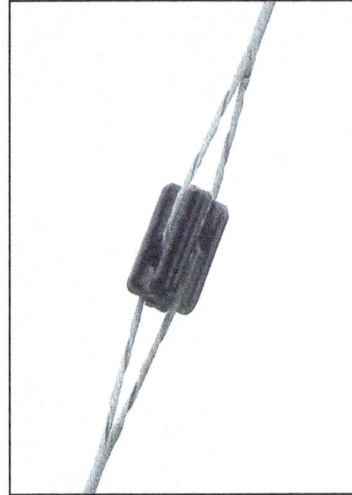

Guy lines have egg-shaped strain insulators in them, to prevent the high voltage on the mast from reaching the ground.

There are three ways of feeding a mast radiator from a transmitter:

- Series excited: The mast is supported on an insulator, and the transmitter is connected to the mast just above it;

- Shunt excited: The mast is grounded and the transmitter feeds it via a wire connected to the mast part way up. (This is a similar approach, on a larger scale, to the 'gamma match' popular among amateur radio operators for VHF and UHF amateur radio antennas.)

- Sectional: The structure is divided into usually two sections with insulators between, usually center-fed. This collinear arrangement enhances low-angle (ground wave) radiation and reduces high-angle (sky wave) radiation. This increases the distance to the mush zone where the ground wave and sky wave are at similar strength at night. This type of antenna is known as an anti-fading aerial. Practical sectionals with 120 over 120 degrees, 180 over 120 degrees and 180 over 180 degrees are presently in operation with good results.

There is usually an antenna matching unit to match the impedance of the transmitter or feeder to the antenna. Depending on the power involved, this may be a small box or a hut or building. It will typically contain an L-network to transform the modulus of impedance, and a coil or capacitor in series with the mast connection to 'tune out' any reactive component.

Location of Transmission Facility

At some facilities, especially the older and higher-powered installations, the mast radiator may be located at a distance from the transmitter building to reduce the field strength induced by the mast into the building, and to prevent the building from distorting the mast's radiation pattern. Between the transmitter building and the antenna matching unit next to the mast radiator, there is a feeder: either an underground coaxial cable or an overhead wire 'cage' feeder.

At facilities with multiple masts, spacings are typically smaller to fit them into the available space.

At modern transmitters or at low power transmitters situated in very small transmitter buildings the transmitter, matching unit and mast radiator can be close together, and possibly in the same building. This saves on feeders, land area and increases the efficiency of the transmitter if only one mast radiator is in use.

At most facilities the mast radiator is on a separate base close to the antenna matching unit, but it can be sometimes be placed on the roof of the antenna matching unit, for example at the main transmission mast of the Mühlacker radio transmitter and the main transmission mast of the Ismaning radio transmitter.

For a good groundwave propagation, mast radiators are built on a large flat area with good ground conductivity, and if possible without inclination. The construction of a mast radiator atop a building or tower whose height is in the same magnitude range as the wavelengths being transmitted gives a bad groundwave propagation. For this reason, mast radiators are (in contrast to FM broadcasting antennas) not typically installed atop buildings or towers. In rare cases, mast radiators for low power transmissions are installed atop buildings. Some lighthouses, such as Reykjanesviti carry a mast radiator for a longwave radio beacon on the roof. The best-known are WGSO in New Orleans and KSBN (AM) on the Delaney Building in Spokane, Washington. In Europe, the low-power broadcasting station at Campobasso uses a mast radiator on a castle.

Fencing

Mast radiators, as with all other equipment showing over 42 volts on exposed components within 4 metres of the ground, are required to be fenced. Usually a chain-link fence is used, but sometimes wooden fences are used to prevent signal interference, which could occur due to currents induced by radio signals in metallic fences. If the mast radiator is mounted atop the antenna tuning hut (which must be over four metres high) or the mast is grounded, with the feed being located not less than four metres above the ground, a fence is not required. It is recommended to fence in any mast radiator to prevent unauthorized climbing.

Ancillary Connections

A mast radiator may need various electrical connections other than the transmitter feed line. Such connections include static drain chokes (which drain off static charges caused by wind, clouds, etc.), spark-gap balls for lightning protection, power supplies for aircraft warning lamps, and coaxial feeders for ancillary antennas mounted on the mast. A variety of techniques are used to 'isolate' these connections from the high RF voltage on the mast, such as chokes, parallel tuned circuits and coupling loops, on a base-fed mast. On a shunt-fed mast, where the base is grounded, such measures are unnecessary.

Anti-fading Antennas

Capacitive "top hat" on mast is occasionally used for electrical lengthening, to allow a
shorter mast to be electrically resonant.

An anti-fading antenna (informally and often incorrectly termed a "Franklin" antenna) is a long-
and medium-wave transmission antenna with a flat vertical radiation pattern. The design goal
is to move the mush zone farther from the transmitter site. An anti-fading antenna will reduce
radiation at elevations of more than 50 degrees as much as possible. In principle, such a radiator
should be as thin as possible, although thicker radiators improve the bandwidth and contribute to
a taller effective height.

These antennas are usually constructed using a metal radio mast, which is fed both at its base and
at an appropriate height. The radio mast is usually insulated from ground and is divided electrically by a separation insulator into two parts. For feeding of the upper part, either a cable inside
the mast construction or the ladder, which must be mounted on insulators, is used. This design
is employed in the Mühlacker, Wolfsheim and Hamburg medium-wave transmitters. Anti-fading
antennas with two separation insulators are installed at Ismaning.

Because separation insulators are fragile compared to the construction of the radio mast framework,
horizontal forces, such as those generated by wind-caused oscillations, need to be minimized. Radio
masts with built-in separation insulators may employ oscillation dampers just above the separation
insulator. This approach is used on the radio masts of Wolfsheim, Hamburg and Ismaning.

A simple anti-fading antenna is a vertical radiator whose height is between 1/2 wavelength (180
electrical degrees) and 5/9 wavelength (200 degrees), 195 degrees being considered ideal. 200 degree and taller simple radiators have largely fallen into disuse, although these can be useful on medium power stations, in which case radiators as tall as 5/8 wavelength (225 degrees) are somewhat
common. Any simple radiator (or array of radiators) which is taller than 195 degrees is generally
not considered to have anti-fading properties.

A more complex anti-fading antenna is a vertical radiator which is sectionalized and is one wavelength (360 degrees, generally 180 over 180 degrees, and historically and correctly termed a
"Franklin" radiator; two in an array at KFBK), or 300 degrees (180 over 120 degrees, and formally
called a WHO radiator), or 240 degrees (120 over 120 degrees, and formally called a WOAI radiator, but WOAI has since replaced this with a 195 degree radiator for lower maintenance cost at

the expense of somewhat poorer "fringe" reception). Several have been incorporated into phased arrays (directional antennas). When so sectionalized the sections are themselves phased so as to provide the optimum vertical radiation pattern (suppressing high-angle radiation). An advantage of the 180 over 180 degree configuration is no ground radial system is required, and the bottom of the lower section may be insulated from the ground; for the others, the bottom of the lower section is usually connected to a ground radial system through a capacitor.

Another form of anti-fading mast antenna is a circular array antenna, in which a number of mast radiators are arranged on a circle and fed in equal phase. With this design, very flat radiation patterns may be produced, although they are very expensive since multiple radio masts are required. Taldom transmitter and Tulagino transmitter in Russia are currently the only radio stations in the world using such antennas. At one time, longwave transmitter Orlunda in Sweden also used an antenna of this type.

HALF-WAVE DIPOLE

A half-wave dipole antenna consists of two quarter-wavelength conductors placed end to end for a total length of approximately L = λ/2.). This is the most widely used antenna because of its advantages. It is also known as Hertz antenna.

Frequency Range

The range of frequency in which half-wave dipole operates is around 3KHz to 300GHz. This is mostly used in radio receivers.

Construction and Working of Half-wave Dipole

It is a normal dipole antenna, where the frequency of its operation is half of its wavelength. Hence, it is called as half-wave dipole antenna.

The edge of the dipole has maximum voltage. This voltage is alternating (AC) in nature. At the positive peak of the voltage, the electrons tend to move in one direction and at the negative peak, the electrons move in the other direction. This can be explained by the figures given below.

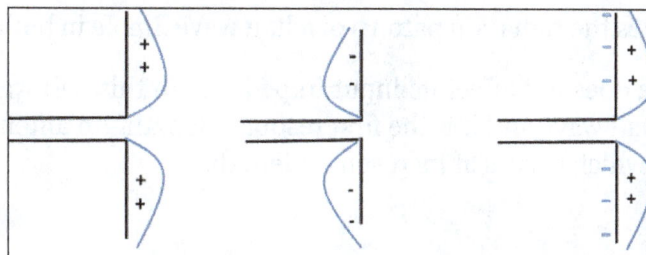

The figures given above show the working of a half-wave dipole.

- First figure shows the dipole when the charges induced are in positive half cycle. Now the electrons tend to move towards the charge.

- Second figure shows the dipole with negative charges induced. The electrons here tend to move away from the dipole.

- Third figure shows the dipole with next positive half cycle. Hence, the electrons again move towards the charge.

The cumulative effect of this produces a varying field effect which gets radiated in the same pattern produced on it. Hence, the output would be an effective radiation following the cycles of the output voltage pattern. Thus, a half-wave dipole radiates effectively.

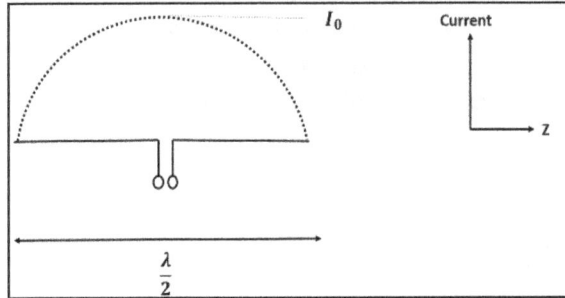

The above figure shows the current distribution in half wave dipole. The directivity of half wave dipole is 2.15dBi, which is reasonably good. Where, 'i' represents the isotropic radiation.

Radiation Pattern

The radiation pattern of this half-wave dipole is Omni-directional in the H-plane. It is desirable for many applications such as mobile communications, radio receivers etc.

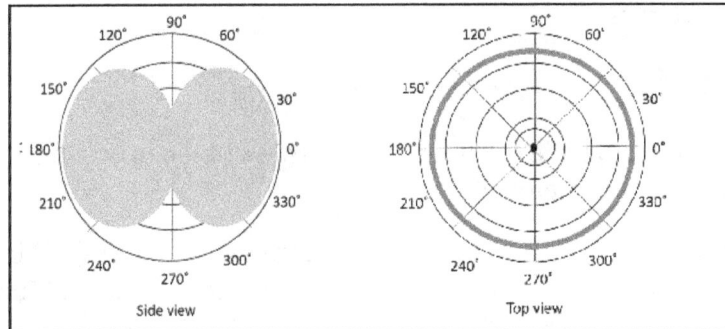

The above figure indicates the radiation pattern of a half wave dipole in both H-plane and V-plane.

The radius of the dipole does not affect its input impedance in this half wave dipole, because the length of this dipole is half wave and it is the first resonant length. An antenna works effectively at its resonant frequency, which occurs at its resonant length.

Advantages

The following are the advantages of half-wave dipole antenna:

- Input impedance is not sensitive.

- Matches well with transmission line impedance.

- Has reasonable length.

- Length of the antenna matches with size and directivity.

Disadvantages

The following are the disadvantages of half-wave dipole antenna:

- Not much effective due to single element.

- It can work better only with a combination.

Applications

The following are the applications of half-wave dipole antenna:

- Used in radio receivers.

- Used in television receivers.

- When employed with others, used for wide variety of applications.

Radiated Fields

Given the length of the dipole, it seems doubtful that the current distribution will be uniform as with the case of the Hertzian dipole. If we think about an open-circuited transmission line made of two wires, we imagine a sinusoidal current distribution set up by the standing wave along a quarter-wavelength length of line as follows:

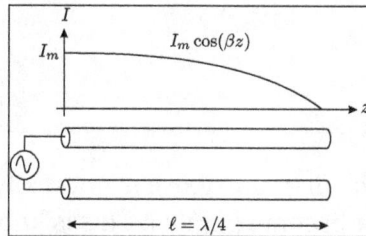

There is no current at $z = \lambda/4$ as required by the open circuit boundary condition. Now, if we "open" up the transmission line, we can essentially create a dipole that is half a wavelength long:

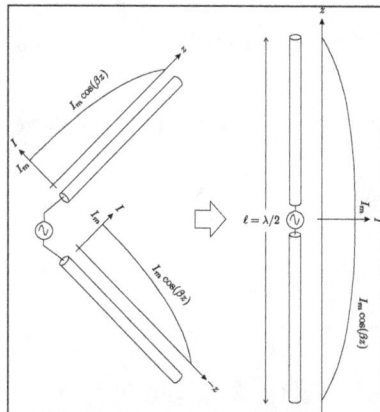

We can write the current distribution as:

$$I(z) = I_m \cos(\beta z)$$

where β is the phase constant associated with the transmission line from which we have drawn the current distribution. Since we are in free space, $\beta = \omega/c = k$. Knowing the current distribution, our next question is how to find the electric field produced by the dipole? Well, we know that a tiny piece of dipole produces an electric field in the far field of,

$$E_\theta = \frac{I\Delta z j\omega\mu}{4\pi}\frac{e^{-jkr}}{r}\sin\theta$$

if excited with a current element of amplitude I at the origin. Using superposition, we can represent the half-wave dipole as a collection of Hertzian dipoles and add up all the responses of each dipole. Hence, each dipole "piece" contributes an electric field,

$$dE_\theta = \frac{I(z')dz'j\omega\mu}{4\pi}\frac{e^{-jkR}}{R}\sin\theta.$$

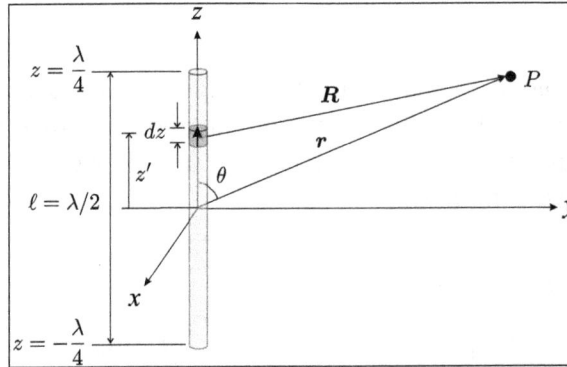

Now, since we are in the far field, the diagram above is not really correct. As the point P moves far from the source, the vectors R and r become parallel. This is known as the parallel ray approximation. Under this approximation,

$$\frac{1}{R} \approx \frac{1}{r}\text{ for amplitude variations}$$

$$\exp(-jkR) \approx \exp\left[-jk(r - z'\cos\theta)\right]\text{ for phase variations}$$

where the latter approximation is evident by examining the geometry of the far-field situation. Then,

$$dE_\theta = \frac{I(z')dz'j\omega\mu}{4\pi}\frac{e^{-jkr}}{r}e^{-jkz'\cos\theta}\sin\theta$$

$$E_\theta = \int_{z'=-\lambda/4}^{z'=\lambda/4}\frac{I(z')dz'j\omega\mu}{4\pi}\frac{e^{-jkr}}{r}e^{-jkz'\cos\theta}\sin\theta dz'$$

$$= \frac{j\omega\mu}{4\pi}\frac{e^{-jkr}}{r}\int_{z'=-\lambda/4}^{z'=\lambda/4}I_m\cos(\beta z')e^{-jkz'\cos\theta}\sin\theta dz'$$

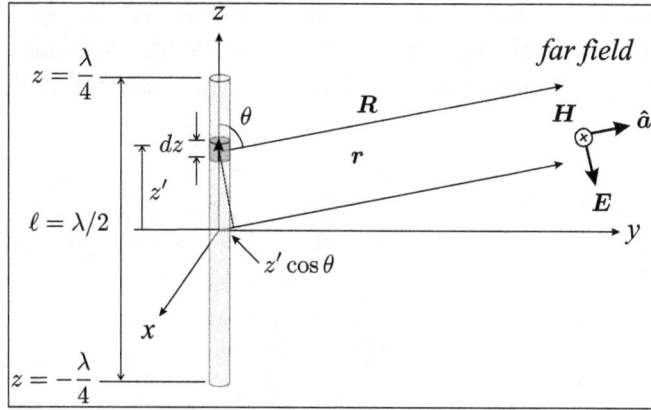

Note:

$$\int \sin(a+bx)e^{cx} = \frac{e^{cx}}{b^2+c^2}\left[c\sin(a+bx) - b\cos(a + bx)\right] + C$$

$$\int_{z'=-\lambda/4}^{z'=\lambda/4} \sin(\pi/2+kz')e^{jkz'\cos\theta} = \frac{e^{jkz'\cos\theta}}{k^2+(jk\cos\theta)^2}\left[jk\cos\theta\sin(\pi/2+kz') - \beta\cos(\pi/2+kz')\right]\Big|_{-\lambda/4}^{\lambda/4}$$

$$= \frac{e^{jk\frac{\lambda}{4}\cos\theta}}{k^2-k^2\cos^2\theta}\left[jk\cos\theta\sin\left(\pi/2 + k\lambda/4\right) - \beta\cos(\pi/2+\beta\lambda/4)\right] -$$

$$\frac{e^{-jk\frac{\lambda}{4}\cos\theta}}{k^2-k^2\cos^2\theta}\left[jk\cos\theta\sin\left(\pi/2 + k\lambda/4\right) - beta\cos(\pi/2+\beta\lambda/4)\right]$$

$$= \frac{e^{j\frac{\pi}{2}\cos\theta}}{\beta^2\sin^2\theta}\beta + \frac{e^{-j\frac{\pi}{2}\cos\theta}}{\beta^2\sin^2\theta}\beta = 2\frac{\cos\left(\frac{\pi}{2}\cos\theta\right)}{\beta\sin^2\theta}$$

Therefore,

$$E_\theta = \underbrace{\frac{j\omega\mu I_m}{4\pi}\frac{e^{-jkr}}{r}\sin\theta}_{\text{Hertzian dipole E-field}} \cdot \underbrace{2\frac{\cos\left(\frac{\pi}{2}\cos\theta\right)}{\beta\sin^2\theta}}_{\text{space factor}}$$

and since $\beta = k$ and $\omega\mu/k = \eta$,

$$E_\theta = \frac{j\eta I_m}{2\pi}\frac{e^{-jkr}}{r}\frac{\cos\left(\frac{\pi}{2}\cos\theta\right)}{\sin\theta}.$$

H_ϕ follows as

$$H_\phi = \frac{E_\theta}{\eta} = \frac{jI_m}{2\pi}\frac{e^{-jkr}}{r}\frac{\cos\left(\frac{\pi}{2}\cos\theta\right)}{\sin\theta}.$$

If we take a polar plot of the pattern indicated by the above expressions, and compare to he pattern from a Hertzian dipole, we notice that a half-wave dipole has slightly less beamwidth then the Hertzian dipole. In fact, the HPBW of a Hertzian dipole is 90∘ , while that of a half-wave dipole is only 78∘ . Hence, we expect the half-wave dipole to exhibit slightly more directivity than its Hertzian counterpart.

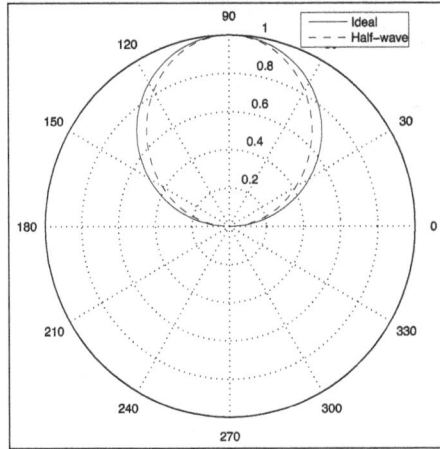

Directivity and Input Impedance

Let's evaluate the directivity and input impedance of the half-wave dipole at the frequency where the dipole is exactly half a wavelength long. We begin by calculating the radiation intensity produced by the dipole:

$$U(\theta) = \frac{1}{2} r^2 \frac{|E_\theta|^2}{\eta} = \frac{1}{2} \frac{\eta I_m^2}{(2\pi)^2} \frac{\cos^2(\pi/2 \cos\theta)}{\sin^2\theta}.$$

The radiated power produced by the dipole is,

$$W_{rad} = \int_0^{2\pi} \int_0^\pi U(\theta) \sin\theta \, d\theta \, d\phi$$

$$= \frac{1}{2}(2\pi) \frac{\eta I_m^2}{(2\pi)^2} \underbrace{\int_0^\pi \frac{\cos^2\left(\frac{\pi}{2}\cos\theta\right)}{\sin\theta} d\theta}_{1.2188 \text{ numerically}}$$

$$= 30(1.2188) I_m^2 = 36.5640 I_m^2.$$

The directivity relative to an isotropic radiator is then calculated as,

$$D_m = \frac{4\pi U_m}{W_{rad}} = \frac{4\pi}{8\pi^2} \eta I_m^2 \cdot \frac{1}{36.5640 I_m^2} = 1.64$$

Therefore,

$$D_{dipole} = 1.64 = 2.15 \, \text{dBi} = 0 \, \text{dBd}.$$

Notice that the dBd unit expresses the directivity with respect to a half-wave dipole, and hence compared to itself, a half-wave dipole has 0 dBd of gain.

For the input impedance, we anticipate both a real and imaginary part, since the near-fields of the dipole will contribute to a reactive component. The input resistance can be found as follows: Then,

$$R_{rad} = \frac{2W_{rad}}{I_m^2} = 73.1280\Omega$$

The calculation of the reactive part of the input impedance is much more involved and beyond the scope of the discussion here. The final result for the dipole's input impedance is

$$Z_{dipole} = 73 + j42.5\ \Omega.$$

That is, the input impedance of the dipole is slightly inductive. However, there exists a "resonance" frequency where the imaginary part of the dipole's input impedance goes to zero. This occurs at a slightly lower frequency and produces,

$$Z_{dipole} = 70 + j0\ \Omega.$$

which is a useful operating point for the antenna. Common coaxial lines, such as RG-59U, have a characteristic impedance of 75 Ω and hence can readily be connected to a dipole without impedance matching, although usually one cannot feed dipoles directly from coaxial line.

Finally, the ohmic loss in a half-wave dipole is,

$$R_{ohmic} = \frac{R_s}{2\pi a}\frac{\lambda}{4}.$$

The details of this calculation have been omitted, but this is not the same expression as a Hertzian dipole. The reason for this is that the ohmic losses are a function of position because the current is not uniformly distributed along the length of the dipole. In fact, if one plugs in L = λ/4 into the expression for the Hertzian dipole and compare to the above expression, the ohmic loss is twice that predicted by equation above, suggesting that only half of the dipole effectively contributes to significant ohmic losses.

References

- Antenna-theory-wire, antenna-theory: tutorialspoint.com, Retrieved 3, May 2019
- poole, ian (2003). Newnes guide to radio and communications technology. Elsevier. Pp. 113–114. Isbn 0-7506-5612-3
- Short dipole, antennas: antenna-theory.com , Retrieved 16, April 2019
- Rick Karlquist (17 Oct 2008). "Low Band Receiving Loops" (PDF). n6rk.com. PacifiCon presentation. Retrieved 2018-04-29
- Antenna-theory-short-dipole, antenna-theory: tutorialspoint.com , Retrieved July 19, 2019
- Das, sisir k. (2016). Antenna and wave propagation. Tata mcgraw-hill education. P. 116. Isbn 1259006328

4

Aperture Antennas

The antenna with an aperture at the edge of a transmission line which radiates energy is known as aperture antenna. A few of its types are slot antenna, parabolic antenna, horn antenna, etc. This chapter closely examines these types of aperture antenna to provide an extensive understanding of the subject.

Aperture antennas constitute a large class of antennas, which emit EM waves through an opening (or aperture). These antennas have close analogs in acoustics, namely, the megaphone and the parabolic microphone. The pupil of the human eye is a typical aperture receiver for optical radiation. At radio and microwave frequencies, horns, waveguide apertures, reflectors and microstrip patches are examples of aperture antennas. Aperture antennas are commonly used at UHF and above where their size is reasonable. Their gain increases as $\sim f^2$. For an aperture antenna to be efficient and to have high directivity, it has to have an area $\geq \lambda^2$ Thus, these antennas are impractical at low frequencies.

To facilitate the analysis of these antennas, the equivalence principle is applied. This allows for carrying out the far-field analysis in the outer (unbounded) region only, which is external to the antenna. This requires the knowledge of the tangential field components at the aperture.

SLOT ANTENNA

Slot radiators or slot antennas are antennas that are used in the frequency range from about 300 MHz to 25 GHz. They are often used in navigation radar usually as an array fed by a waveguide. But also older large phased array antennas used the principle because the slot radiators are a very inexpensive way for frequency scanning arrays. Slot antennas are an about $\lambda/2$ elongated slot, cut in a conductive plate (Consider an infinite conducting sheet), and excited in the center. This slot behaves according to Babinet's principle as resonant radiator. Jacques Babinet was a French physicist and mathematician, formulated the theorem that similar diffraction patterns are produced by two complementary screens (Babinet's principle). This principle relates the radiated fields and impedance of an aperture or slot antenna to that of the field of a dipole antenna. The polarization of a slot antenna is linear. The fields of the slot antenna are almost the same as the dipole antenna, but the field's components are interchanged: a vertical slot has got an horizontal electric field; and the vertical dipole has got a vertical electrical field.

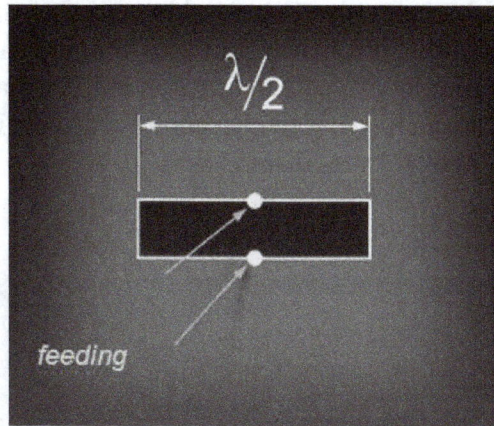

The length of a slot determines the resonant frequency, the width of the slit
determines the broad bandwidth of the slot radiator.

The impedance of the slot antenna (Zs) is related to the impedance of its complementary dipole
antenna (Zd) by the relation:

$$Zd \cdot Zs = \eta^2/4 \qquad \text{mit } Zs = \text{impedance of the slot antenna}$$

Zd = impedance of its dual antenna

η = intrinsic impedance of free space. It follows for Zs = 485 Ω.

The band width of a narrow rectangular slot is equal to that of the related dipole, and is equal
to half the bandwidth of a cylindrical dipole with a diameter equal to the slot width. Figure
shows slot antennas different from the rectangular shape that increasing the bandwidth of the
slot antenna.

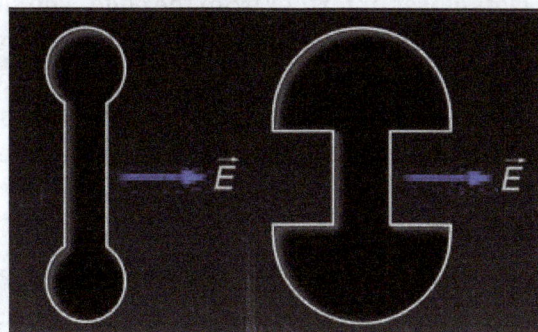

Various broadband slot antenna.

Although the theory requires an infinite spread conductive surface, the deviation from the theoretical value is small when the surface is greater than the square of the wavelength. The feeding of the
slot antenna can be done with ordinary two-wire line. The impedance is dependent on the feeding
point, as in a dipole. The value of 485 Ω applies only to a feeding point at the center. A shift of the
feed point from the center to the edge steadily decreases the impedance.

The application of slot antennas can be versatile. They can replace dipoles e.g. if it is required a polarization perpendicular to the longitudinal extension of the radiator. If a dipole is used for feeding
of a parabolic antenna to generate a vertically orientated but horizontally polarized fan beam, then

this dipole must be orientated horizontally. This would mean that the edge surfaces of the parabolic reflector will not be sufficiently illuminated, but a lot of energy above and below the reflector would be lost. In addition, the length of the dipole is extended in a plane, in which is demanding a point like source of radiation for the focus of the parabolic reflector. If this dipole is replaced by a slot antenna, in this case don't appear these disadvantages.

Slots in Waveguides

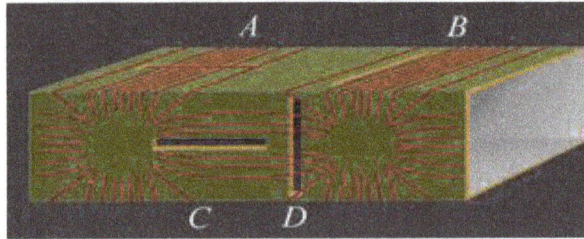

Various slot arrangements in a waveguide.

Slot antennas in waveguides provide an economical way of the design of antenna arrays. The position, shape and orientation of the slots will determine how (or if) they radiate. Figure shows a rectangular waveguide with a drawn with red lines snapshot of the schematic current distribution in the waveguide walls. If slots are cut into the walls, so the current flow is affected more or less depending on the location of the slot. If the slots are sufficiently narrow so the slots B and C have little influence on the current distribution. These two slots radiate not (or very little). The slots A and D represent barriers to the current flow. Thus, this current flow acts as an excitation system for the slot, this one acts as radiator. Since the wave in the waveguide moves forward, these drawn lines migrate in the direction of propagation. The slot gets one always alternating voltage potential at its slot edges (depending on the frequency in the waveguide). The power that the slot radiates can be altered by moving the slots closer or farther from the edge. The slots A and D have the strongest coupling to the RF energy transported in the waveguide. In order to reduce this coupling, for example the slot A could be moved closer to one of the shorter waveguide walls. Rotating of the slots would have a the same effect (an angle between the orientations of A and B or C and D). The coupling of this rotated slot ist a factor of about sin2 of the rotating angle θ.

Slotted Waveguide Antennas

Basic geometry of a slotted waveguide antenna. (The slot radiators are on the wider wall of the rectangular waveguide).

Several slot radiators in a waveguide form a group antenna. The waveguide is used as the transmission line to feed the elements. In order for radiate in the correct phase, all single slots must be cutted in the distance of the wavelength, that is valid for the interior of the waveguide. This wavelength differs from the wavelength in free space and is a function of the wider side a of a rectangular waveguide. Usually this wavelength is calculated for the TE_{10} mode by:

$$\lambda_h = \frac{\lambda}{\sqrt{1 - \left(\dfrac{\lambda}{2a}\right)^2}}$$

- a = length of the wider side of the rectangular waveguides

- λh = "guided" wavelength (within the waveguide)

- λ = wavelength in free space

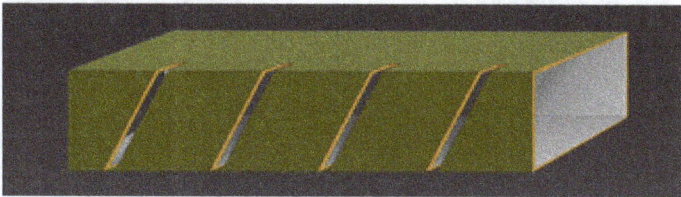

Basic geometry of a slotted waveguide antenna with
rotated slot antennas on the narrower wall.

The wavelength within the waveguide is longer than in free space. The distance of the slot radiators in the group is set at this wavelength to a value that is slightly larger than the wavelength λ in the free space. The number and the size of the side lobes is affected so unfavorably. The slots are often attached to the left and right eccentrically (with reduced coupling). If mounted on the narrow side of the waveguide, it may happen that the length for the resonant slot radiator is shorter than the wall. In this case, the slot can be also guided around the corners, it then lies also slightly on the Aside of the waveguide. In practice, these slots are all covered with a thin insulating material (for the protection of the interior) of the waveguide. This material may not be hygroscopic and must be protected from weather conditions.

A single narrow slot radiator can also work on frequencies ±5 ... ±10% besides its resonance frequency. For array antennas, this is not possible so easily. Such a group antenna is fixed strongly to a single frequency, which is determined by the spacing of exactly λg, and for which the antenna has been optimized. If the frequency is changed, then these distances not correct, the performance of the antenna decreases. The phase difference arising between the antenna elements are added to the whole length of the antenna to values that can no longer be tolerated. This antenna begins to "squint", that is, the antenna pattern points in a different direction from the optical center axis. This effect can also be exploited to achieve an electronic pivoting of the antenna beam as a function of change of the transmission frequency.

PARABOLIC ANTENNA

A parabolic antenna is an antenna that uses a parabolic reflector, a curved surface with the cross-sectional shape of a parabola, to direct the radio waves. The most common form is shaped like a dish and is popularly called a dish antenna or parabolic dish. The main advantage of a parabolic antenna is that it has high directivity. It functions similarly to a searchlight or flashlight reflector to direct the radio waves in a narrow beam, or receive radio waves from one particular direction only. Parabolic antennas have some of the highest gains, meaning that they can produce the narrowest beamwidths, of any antenna type. In order to achieve narrow beamwidths, the parabolic reflector must be much larger than the wavelength of the radio waves used, so parabolic antennas are used in the high frequency part of the radio spectrum, at UHF and microwave (SHF) frequencies, at which the wavelengths are small enough that conveniently-sized reflectors can be used.

A large parabolic satellite communications antenna at Erdfunkstelle Raisting, the biggest facility for satellite communication in the world, in Raisting, Bavaria, Germany. It has a Cassegrain type feed.

Parabolic antennas are used as high-gain antennas for point-to-point communications, in applications such as microwave relay links that carry telephone and television signals between nearby cities, wireless WAN/LAN links for data communications, satellite communications and spacecraft communication antennas. They are also used in radio telescopes.

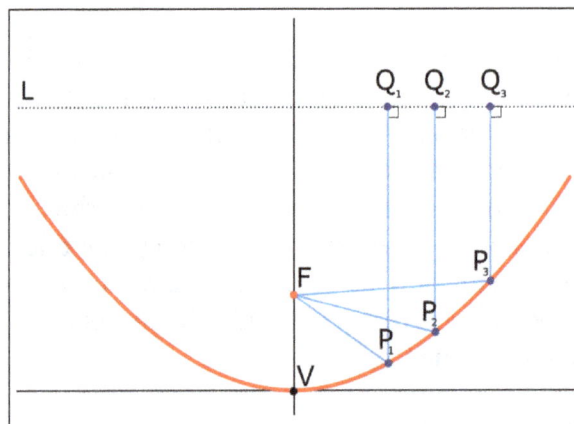

Parabolic antennas are based on the geometrical property of the paraboloid that the paths FP_1Q_1, FP_2Q_2, FP_3Q_3 are all the same length. So a spherical wavefront emitted by a feed antenna at the dish's focus F will be reflected into an outgoing plane wave L travelling parallel to the dish's axis VF.

The other large use of parabolic antennas is for radar antennas, in which there is a need to transmit a narrow beam of radio waves to locate objects like ships, airplanes, and guided missiles, and often

Aperture Antennas

101

for weather detection. With the advent of home satellite television receivers, parabolic antennas have become a common feature of the landscapes of modern countries.

The parabolic antenna was invented by German physicist Heinrich Hertz during his discovery of radio waves in 1887. He used cylindrical parabolic reflectors with spark-excited dipole antennas at their focus for both transmitting and receiving during his historic experiments.

Design

The operating principle of a parabolic antenna is that a point source of radio waves at the focal point in front of a paraboloidal reflector of conductive material will be reflected into a collimated plane wave beam along the axis of the reflector. Conversely, an incoming plane wave parallel to the axis will be focused to a point at the focal point.

A typical parabolic antenna consists of a metal parabolic reflector with a small feed antenna suspended in front of the reflector at its focus, pointed back toward the reflector. The reflector is a metallic surface formed into a paraboloid of revolution and usually truncated in a circular rim that forms the diameter of the antenna. In a transmitting antenna, radio frequency current from a transmitter is supplied through a transmission line cable to the feed antenna, which converts it into radio waves. The radio waves are emitted back toward the dish by the feed antenna and reflect off the dish into a parallel beam. In a receiving antenna the incoming radio waves bounce off the dish and are focused to a point at the feed antenna, which converts them to electric currents which travel through a transmission line to the radio receiver.

Parabolic Reflector

Wire grid-type parabolic antenna used for MMDS data link at a frequency of 2.5-2.7 GHz. It is fed by a vertical dipole under the small aluminum reflector on the boom. It radiates vertically polarized microwaves.

The reflector can be of sheet metal, metal screen, or wire grill construction, and it can be either a circular "dish" or various other shapes to create different beam shapes. A metal screen reflects radio waves as well as a solid metal surface as long as the holes are smaller than one-tenth of a wavelength, so screen reflectors are often used to reduce weight and wind loads on the dish. To achieve the maximum gain, it is necessary that the shape of the dish be accurate within a small fraction of a wavelength, to ensure the waves from different parts of the antenna arrive at the focus in phase. Large dishes often require a supporting truss structure behind them to provide the required stiffness.

A reflector made of a grill of parallel wires or bars oriented in one direction acts as a *polarizing filter* as well as a reflector. It only reflects linearly polarized radio waves, with the electric field parallel to the grill elements. This type is often used in radar antennas. Combined with a linearly polarized feed horn, it helps filter out noise in the receiver and reduces false returns.

Since a shiny metal parabolic reflector can also focus the sun's rays, and most dishes could concentrate enough solar energy on the feed structure to severely overheat it if they happened to be pointed at the sun, solid reflectors are always given a coat of flat paint.

Feed Antenna

The feed antenna at the reflector's focus is typically a low-gain type such as a half-wave dipole or more often a small horn antenna called a feed horn. In more complex designs, such as the Cassegrain and Gregorian, a secondary reflector is used to direct the energy into the parabolic reflector from a feed antenna located away from the primary focal point. The feed antenna is connected to the associated radio-frequency (RF) transmitting or receiving equipment by means of a coaxial cable transmission line or waveguide.

At the microwave frequencies used in many parabolic antennas, waveguide is required to conduct the microwaves between the feed antenna and transmitter or receiver. Because of the high cost of waveguide runs, in many parabolic antennas the RF front end electronics of the receiver is located at the feed antenna, and the received signal is converted to a lower intermediate frequency (IF) so it can be conducted to the receiver through cheaper coaxial cable. This is called a low-noise block downconverter. Similarly, in transmitting dishes, the microwave transmitter may be located at the feed point.

An advantage of parabolic antennas is that most of the structure of the antenna (all of it except the feed antenna) is nonresonant, so it can function over a wide range of frequencies, that is a wide bandwidth. All that is necessary to change the frequency of operation is to replace the feed antenna with one that works at the new frequency. Some parabolic antennas transmit or receive at multiple frequencies by having several feed antennas mounted at the focal point, close together.

Dish Parabolic Antennas

Shrouded microwave relay dishes on a communications tower in Australia.

A satellite television dish, an example of an offset fed dish.

Cassegrain satellite communication antenna in Sweden.

Offset Gregorian antenna used in the Allen Telescope Array, a radio telescope at the University of California Berkeley, US.

Shaped-beam Parabolic Antennas

Vertical "orange peel" antenna for military height finder radar, Germany.

Air traffic control radar antenna, near Hannover, Germany.

ASR-9 Airport surveillance radar antenna.

Types

Parabolic antennas are distinguished by their shapes:

- Paraboloidal or dish – The reflector is shaped like a paraboloid truncated in a circular rim. This is the most common type. It radiates a narrow pencil-shaped beam along the axis of the dish.

  ◦ Shrouded dish – Sometimes a cylindrical metal shield is attached to the rim of the dish. The shroud shields the antenna from radiation from angles outside the main beam axis, reducing the sidelobes. It is sometimes used to prevent interference in terrestrial microwave links, where several antennas using the same frequency are located close together. The shroud is coated inside with microwave absorbent material. Shrouds can reduce back lobe radiation by 10 dB.

- Cylindrical – The reflector is curved in only one direction and flat in the other. The radio waves come to a focus not at a point but along a line. The feed is sometimes a dipole antenna located along the focal line. Cylindrical parabolic antennas radiate a fan-shaped beam, narrow in the curved dimension, and wide in the uncurved dimension. The curved ends of the reflector are sometimes capped by flat plates, to prevent radiation out the ends, and this is called a *pillbox* antenna.

- Shaped-beam antennas – Modern reflector antennas can be designed to produce a beam or beams of a particular shape, rather than just the narrow "pencil" or "fan" beams of the simple dish and cylindrical antennas above. Two techniques are used, often in combination, to control the shape of the beam:

  ◦ Shaped reflectors – The parabolic reflector can be given a noncircular shape, and/ or different curvatures in the horizontal and vertical directions, to alter the shape of the beam. This is often used in radar antennas. As a general principle, the wider the antenna is in a given transverse direction, the narrower the radiation pattern will be in that direction.

◻ "Orange peel" antenna – Used in search radars, this is a long narrow antenna shaped like the letter "C". It radiates a narrow vertical fan shaped beam.

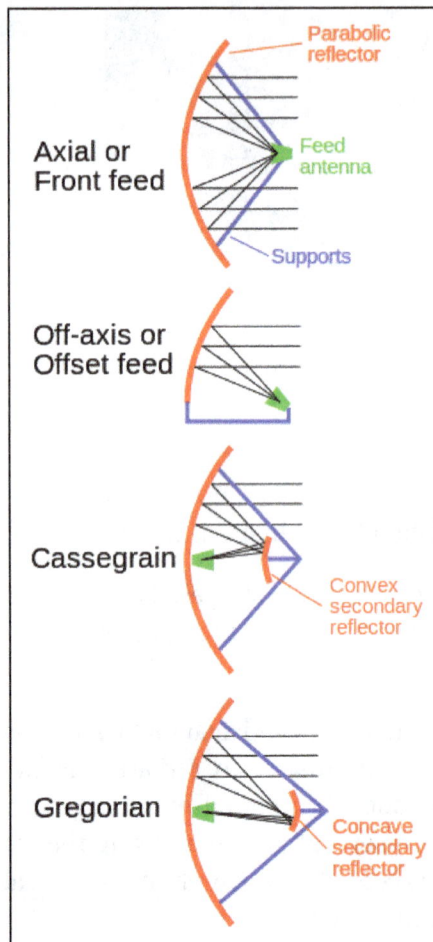

Main types of parabolic antenna feeds.

- Arrays of feeds – In order to produce an arbitrary shaped beam, instead of one feed horn, an array of feed horns clustered around the focal point can be used. Array-fed antennas are often used on communication satellites, particularly direct broadcast satellites, to create a downlink radiation pattern to cover a particular continent or coverage area. They are often used with secondary reflector antennas such as the Cassegrain.

Array of multiple feed horns on a German airport surveillance radar antenna to control the elevation angle of the beam.

Parabolic antennas are also classified by the type of *feed*, that is, how the radio waves are supplied to the antenna:

- *Axial*, *prime focus*, or *front feed* – This is the most common type of feed, with the feed antenna located in front of the dish at the focus, on the beam axis, pointed back toward the dish. A disadvantage of this type is that the feed and its supports block some of the beam, which limits the aperture efficiency to only 55–60%.

- *Off-axis* or *offset feed* – The reflector is an asymmetrical segment of a paraboloid, so the focus, and the feed antenna, are located to one side of the dish. The purpose of this design is to move the feed structure out of the beam path, so it does not block the beam. It is widely used in home satellite television dishes, which are small enough that the feed structure would otherwise block a significant percentage of the signal. Offset feed can also be used in multiple reflector designs such as the Cassegrain and Gregorian, below.

- *Cassegrain* – In a Cassegrain antenna, the feed is located on or behind the dish, and radiates forward, illuminating a convex hyperboloidal secondary reflector at the focus of the dish. The radio waves from the feed reflect back off the secondary reflector to the dish, which reflects them forward again, forming the outgoing beam. An advantage of this configuration is that the feed, with its waveguides and "front end" electronics does not have to be suspended in front of the dish, so it is used for antennas with complicated or bulky feeds, such as large satellite communication antennas and radio telescopes. Aperture efficiency is on the order of 65–70%

- *Gregorian* – Similar to the Cassegrain design except that the secondary reflector is concave, (ellipsoidal) in shape. Aperture efficiency over 70% can be achieved.

Feed Pattern

The radiation pattern of the feed antenna has to be tailored to the shape of the dish, because it has a strong influence on the *aperture efficiency*, which determines the antenna gain. Radiation from the feed that falls outside the edge of the dish is called "*spillover*" and is wasted, reducing the gain and increasing the backlobes, possibly causing interference or (in receiving antennas) increasing susceptibility to ground noise. However, maximum gain is only achieved when the dish is uniformly "illuminated" with a constant field strength to its edges. So the ideal radiation pattern of a feed antenna would be a constant field strength throughout the solid angle of the dish, dropping abruptly to zero at the edges. However, practical feed antennas have radiation patterns that drop off gradually at the edges, so the feed antenna is a compromise between acceptably low spillover and adequate illumination. For most front feed horns, optimum illumination is achieved when the power radiated by the feed horn is 10 dB less at the dish edge than its maximum value at the center of the dish.

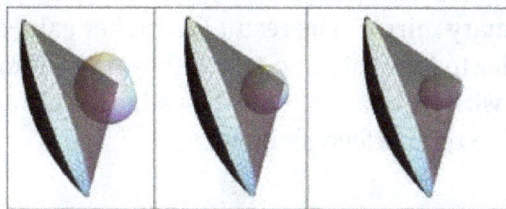

Effect of the feed antenna radiation pattern *(small pumpkin-shaped surface)* on spillover.
Left: With a low gain feed antenna, significant parts of its radiation fall outside the dish.
Right: With a higher gain feed, almost all its radiation is emitted within the angle of the dish.

Polarization

The pattern of electric and magnetic fields at the mouth of a parabolic antenna is simply a scaled up image of the fields radiated by the feed antenna, so the polarization is determined by the feed antenna. In order to achieve maximum gain, the feed antenna in the transmitting and receiving antenna must have the same polarization. For example, a vertical dipole feed antenna will radiate a beam of radio waves with their electric field vertical, called vertical polarization. The receiving feed antenna must also have vertical polarization to receive them; if the feed is horizontal (horizontal polarization) the antenna will suffer a severe loss of gain.

To increase the data rate, some parabolic antennas transmit two separate radio channels on the same frequency with orthogonal polarizations, using separate feed antennas; this is called a *dual polarization* antenna. For example, satellite television signals are transmitted from the satellite on two separate channels at the same frequency using right and left circular polarization. In a home satellite dish, these are received by two small monopole antennas in the feed horn, oriented at right angles. Each antenna is connected to a separate receiver.

If the signal from one polarization channel is received by the oppositely polarized antenna, it will cause crosstalk that degrades the signal-to-noise ratio. The ability of an antenna to keep these orthogonal channels separate is measured by a parameter called *cross polarization discrimination* (XPD). In a transmitting antenna, XPD is the fraction of power from an antenna of one polarization radiated in the other polarization. For example, due to minor imperfections a dish with a vertically polarized feed antenna will radiate a small amount of its power in horizontal polarization; this fraction is the XPD. In a receiving antenna, the XPD is the ratio of signal power received of the opposite polarization to power received in the same antenna of the correct polarization, when the antenna is illuminated by two orthogonally polarized radio waves of equal power. If the antenna system has inadequate XPD, cross polarization interference cancelling (XPIC) digital signal processing algorithms can often be used to decrease crosstalk.

Dual Reflector Shaping

In the Cassegrain and Gregorian antennas, the presence of two reflecting surfaces in the signal path offers additional possibilities for improving performance. When the highest performance is required, a technique called "dual reflector shaping" may be used. This involves changing the shape of the sub-reflector to direct more signal power to outer areas of the dish, to map the known pattern of the feed into a uniform illumination of the primary, to maximize the gain. However, this results in a secondary that is no longer precisely hyperbolic (though it is still very close), so the constant phase property is lost. This phase error, however, can be compensated for by slightly tweaking the shape of the primary mirror. The result is a higher gain, or gain/spillover ratio, at the cost of surfaces that are trickier to fabricate and test. Other dish illumination patterns can also be synthesized, such as patterns with high taper at the dish edge for ultra-low spillover sidelobes, and patterns with a central "hole" to reduce feed shadowing.

Gain

The directive qualities of an antenna are measured by a dimensionless parameter called its gain, which is the ratio of the power received by the antenna from a source along its beam axis to the power received by a hypothetical isotropic antenna. This is,

$$G = \frac{A_{\text{antenna}}}{A_{\text{isotropic}}}$$

The aperture A_{antenna} of the antenna is equal to the area of the physical aperture A multiplied by a factor e_A between 0 and 1 called the aperture efficiency: $A_{\text{antenna}} = e_A A$. The aperture of an isotropic antenna is,

$$A_{\text{isotropic}} = \frac{\lambda^2}{4\pi}$$

Thus the gain of a parabolic antenna is:

$$G = \frac{4\pi A}{\lambda^2} e_A = \left(\frac{\pi d}{\lambda} \right)^2 e_A$$

where:

- A is the area of the antenna aperture, that is, the mouth of the parabolic reflector. For a circular dish antenna, $A = \pi d^2 / 4$, giving the second formula above.

- d is the diameter of the parabolic reflector, if it is circular

- λ is the wavelength of the radio waves.

- e_A is a dimensionless parameter between 0 and 1 called the *aperture efficiency*. The aperture efficiency of typical parabolic antennas is 0.55 to 0.70.

Arecibo radio telescope, in Puerto Rico, US. At 1000 ft (300 m) in diameter, it is the second largest "dish" antenna in the world. The reflector's shape is actually spherical, not paraboloidal, to reduce aberration when it is focused off axis.

It can be seen that, as with any *aperture antenna*, the larger the aperture is, compared to the wavelength, the higher the gain. The gain increases with the square of the ratio of aperture width to wavelength, so large parabolic antennas, such as those used for spacecraft communication and radio telescopes, can have extremely high gain. Applying the above formula to the 25-meter-diameter antennas often used in radio telescope arrays and satellite ground antennas at a wavelength of 21 cm (1.42 GHz, a common radio astronomy frequency), yields an approximate maximum gain of 140,000 times or about 50 dBi (decibels above the isotropic level). The largest parabolic dish antennas in the world are the Five-hundred-meter Aperture Spherical radio Telescope in southwest China, and the Arecibo radio telescope in Arecibo, Puerto Rico, US, which both have effective apertures of about 300 meters. The gain of these dishes at 3 GHz is roughly 90 million, or 80 dBi.

Aperture efficiency e_A is a catchall variable which accounts for various losses that reduce the gain of the antenna from the maximum that could be achieved with the given aperture. The major factors reducing the aperture efficiency in parabolic antennas are:.

- Feed spillover - Some of the radiation from the feed antenna falls outside the edge of the dish and so doesn't contribute to the main beam.

- Feed illumination taper - The maximum gain for any aperture antenna is only achieved when the intensity of the radiated beam is constant across the entire aperture area. However the radiation pattern from the feed antenna usually tapers off toward the outer part of the dish, so the outer parts of the dish are "illuminated" with a lower intensity of radiation. Even if the feed provided constant illumination across the angle subtended by the dish, the outer parts of the dish are farther away from the feed antenna than the inner parts, so the intensity would drop off with distance from the center. So the intensity of the beam radiated by a parabolic antenna is maximum at the center of the dish and falls off with distance from the axis, reducing the efficiency.

- Aperture blockage - In front-fed parabolic dishes where the feed antenna is located in front of the dish in the beam path (and in Cassegrain and Gregorian designs as well), the feed structure and its supports block some of the beam. In small dishes such as home satellite dishes, where the size of the feed structure is comparable with the size of the dish, this can seriously reduce the antenna gain. To prevent this problem these types of antennas often use an *offset* feed, where the feed antenna is located to one side, outside the beam area. The aperture efficiency for these types of antennas can reach 0.7 to 0.8.

- Shape errors - random surface errors in the shape of the reflector reduce efficiency. The loss is approximated by Ruze's Equation.

For theoretical considerations of mutual interference (at frequencies between 2 and c. 30 GHz - typically in the Fixed Satellite Service) where specific antenna performance has not been defined, a reference antenna based on Recommendation ITU-R S.465 is used to calculate the interference, which will include the likely sidelobes for off-axis effects.

Radiation Pattern

In the above figure, the main lobe *(top)* is only a few degrees wide. The sidelobes are all at least 20 dB below (1/100 the power density of) the main lobe, and most are 30 dB below. (If this pattern was drawn with linear power levels instead of logarithmic dB levels, all lobes other than the main lobe would be much too small to see.)

In parabolic antennas, virtually all the power radiated is concentrated in a narrow main lobe along the antenna's axis. The residual power is radiated in sidelobes, usually much smaller, in other directions. Because in parabolic antennas the reflector aperture is much larger than the wavelength, due to diffraction there are usually many narrow sidelobes, so the sidelobe pattern is complex. There is also usually a backlobe, in the opposite direction to the main lobe, due to the spillover radiation from the feed antenna that misses the reflector.

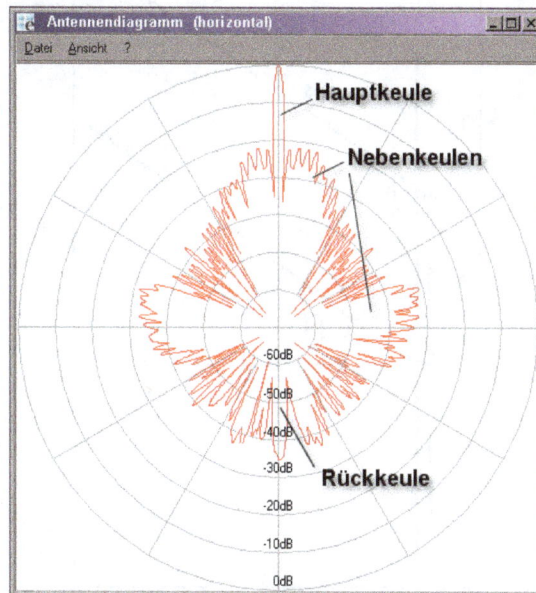

Radiation pattern of a German parabolic antenna.

Beamwidth

The angular width of the beam radiated by high-gain antennas is measured by the *half-power beam width* (HPBW), which is the angular separation between the points on the antenna radiation pattern at which the power drops to one-half (-3 dB) its maximum value. For parabolic antennas, the HPBW θ is given by:

$$\theta = k\lambda / d$$

where, k is a factor which varies slightly depending on the shape of the reflector and the feed illumination pattern. For an ideal uniformly illuminated parabolic reflector and θ in degrees, k would be 57.3 (the number of degrees in a radian). For a "typical" parabolic antenna k is approximately 70.

For a typical 2 meter satellite dish operating on C band (4 GHz), this formula gives a beamwidth of about 2.6°. For the Arecibo antenna at 2.4 GHz the beamwidth is 0.028°. It can be seen that parabolic antennas can produce very narrow beams, and aiming them can be a problem. Some parabolic dishes are equipped with a boresight so they can be aimed accurately at the other antenna.

It can be seen there is an inverse relation between gain and beam width. By combining the beamwidth equation with the gain equation, the relation is:

$$G = \left(\frac{\pi k}{\theta}\right)^2 e_A.$$

Radiation Pattern Formula

The radiation from a large paraboloid with uniform illuminated aperture is essentially equivalent to that from a circular aperture of the same diameter D in an infinite metal plate with a uniform plane wave incident on the plate.

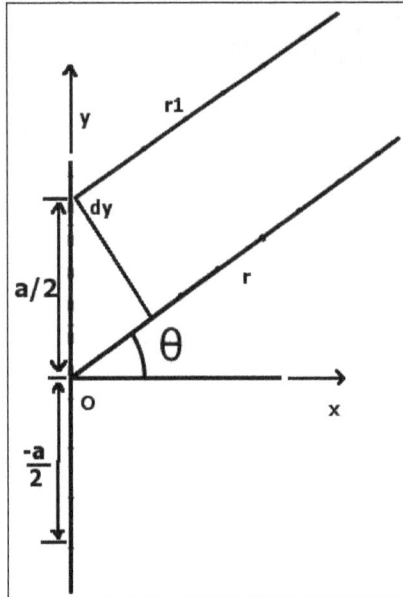

The angle theta is normal to the aperture.

The radiation-field pattern can be calculated by applying Huygens' principle in a similar way to a rectangular aperture. The electric field pattern can be found by evaluating the Fraunhofer diffraction integral over the circular aperture. It can also be determined through Fresnel zone equations.

$$E = \iint -e^{j(\omega t - \beta r)} dS = \iint e^{2\pi i(lx + my)/\lambda} dS$$

where, $\beta = \omega / c = 2\pi / \lambda$. Using polar coordinates $x = \rho \cdot \cos\theta, \quad y = \rho \cdot \sin\theta$. Taking account of symmetry,

$$E = \int_0^{2\pi} d\theta \int_0^{\rho_0} e^{2\pi i \rho \cos\theta l / \lambda} \rho d\rho$$

and using first-order Bessel function gives the electric field pattern $E(\theta)$,

$$E(\theta) = \frac{2\lambda}{\pi D} \frac{J_1[(\pi D / \lambda)\sin\theta]}{\sin\theta}$$

where D is the diameter of the antenna's aperture in meters, λ is the wavelength in meters, θ is the angle in radians from the antenna's symmetry axis as shown in the figure, and J_1 is the first-order Bessel function. Determining the first nulls of the radiation pattern gives the beamwidth θ_0. The term $J_1(x) = 0$ whenever $S(\mathbf{x})$.

Thus,

$$\theta_0 = \arcsin\frac{3.83\lambda}{\pi D} = \arcsin\frac{1.22\lambda}{D}.$$

When the aperture is large the angle θ_0 is very small, so $\arcsin(x)$ is approximataly equal to x This gives the common beamwidth formulas,

$$\theta_0 \approx \frac{1.22\lambda}{D}\text{(in radians)} = \frac{70\lambda}{D}\text{(in degrees).}$$

HORN ANTENNA

A horn antenna or microwave horn is an antenna that consists of a flaring metal waveguide shaped like a horn to direct radio waves in a beam. Horns are widely used as antennas at UHF and microwave frequencies, above 300 MHz. They are used as feed antennas (called feed horns) for larger antenna structures such as parabolic antennas, as standard calibration antennas to measure the gain of other antennas, and as directive antennas for such devices as radar guns, automatic door openers, and microwave radiometers. Their advantages are moderate directivity, low standing wave ratio (SWR), broad bandwidth, and simple construction and adjustment.

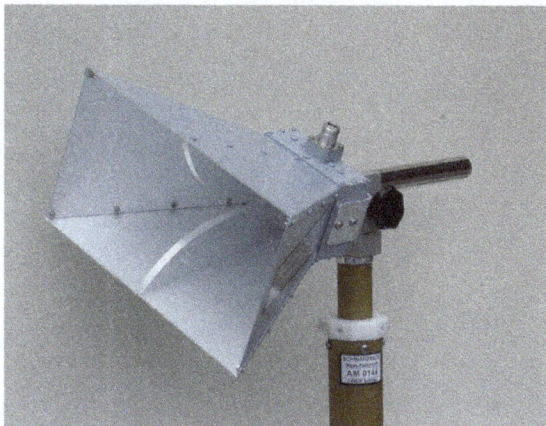

Pyramidal microwave horn antenna, with a bandwidth of 0.8 to 18 GHz. A coaxial cable feedline attaches to the connector visible at top. This type is called a ridged horn; the curving fins visible inside the mouth of the horn increase the antenna's bandwidth.

One of the first horn antennas was constructed in 1897 by Bengali-Indian radio researcher Jagadish Chandra Bose in his pioneering experiments with microwaves. The modern horn antenna was invented independently in 1938 by Wilmer Barrow and G. C. Southworth The development of radar in World War 2 stimulated horn research to design feed horns for radar antennas. The corrugated horn invented by Kay in 1962 has become widely used as a feed horn for microwave antennas such as satellite dishes and radio telescopes.

An advantage of horn antennas is that since they have no resonant elements, they can operate over a wide range of frequencies, a wide bandwidth. The usable bandwidth of horn antennas is typically of the order of 10:1, and can be up to 20:1 (for example allowing it to operate from 1 GHz to 20 GHz). The input impedance is slowly varying over this wide frequency range, allowing low voltage standing wave ratio (VSWR) over the bandwidth. The gain of horn antennas ranges up to 25 dBi, with 10 - 20 dBi being typical.

A horn antenna is used to transmit radio waves from a waveguide (a metal pipe used to carry radio waves) out into space, or collect radio waves into a waveguide for reception. It typically consists of a short length of rectangular or cylindrical metal tube (the waveguide), closed at one end, flaring into an open-ended conical or pyramidal shaped horn on the other end. The radio waves are usually introduced into the waveguide by a coaxial cable attached to the side, with the central conductor projecting into the waveguide to form a quarter-wave monopole antenna. The waves then radiate out the horn end in a narrow beam. In some equipment the radio waves are conducted between the transmitter or receiver and the antenna by a waveguide; in this case the horn is attached to the end of the waveguide. In outdoor horns, such as the feed horns of satellite dishes, the open mouth of the horn is often covered by a plastic sheet transparent to radio waves, to exclude moisture.

How it Works

A horn antenna serves the same function for electromagnetic waves that an acoustical horn does for sound waves in a musical instrument such as a trumpet. It provides a gradual transition structure to match the impedance of a tube to the impedance of free space, enabling the waves from the tube to radiate efficiently into space.

Corrugated conical horn antenna used as a feed horn on a Hughes Direcway home satellite dish. A transparent plastic sheet covers the horn mouth to keep out rain.

If a simple open-ended waveguide is used as an antenna, without the horn, the sudden end of the conductive walls causes an abrupt impedance change at the aperture, from the wave impedance in the waveguide to the impedance of free space, (about 377 ohms). When radio waves travelling through the waveguide hit the opening, this impedance-step reflects a significant fraction of the wave energy back down the guide toward the source, so that not all of the power is radiated. This is similar to the reflection at an open-ended transmission line or a boundary between optical mediums with a low and high index of refraction, like at a glass surface. The reflected waves cause standing waves in the waveguide, increasing the SWR, wasting energy and possibly overheating the transmitter. In addition, the small aperture of the waveguide (less than one wavelength) causes significant diffraction of the waves issuing from it, resulting in a wide radiation pattern without much directivity.

To improve these poor characteristics, the ends of the waveguide are flared out to form a horn. The taper of the horn changes the impedance gradually along the horn's length. This acts like an impedance matching transformer, allowing most of the wave energy to radiate out the end of the horn into space, with minimal reflection. The taper functions similarly to a tapered transmission line, or an optical medium with a smoothly varying refractive index. In addition, the wide aperture of the horn projects the waves in a narrow beam.

The horn shape that gives minimum reflected power is an exponential taper. Exponential horns are used in special applications that require minimum signal loss, such as satellite antennas and radio telescopes. However conical and pyramidal horns are most widely used, because they have straight sides and are easier to design and fabricate.

Radiation Pattern

The waves travel down a horn as spherical wavefronts, with their origin at the apex of the horn, a point called the phase center. The pattern of electric and magnetic fields at the aperture plane at the mouth of the horn, which determines the radiation pattern, is a scaled-up reproduction of the fields in the waveguide. Because the wavefronts are spherical, the phase increases smoothly from the edges of the aperture plane to the center, because of the difference in length of the center point and the edge points from the apex point. The difference in phase between the center point and the edges is called the *phase error*. This phase error, which increases with the flare angle, reduces the gain and increases the beamwidth, giving horns wider beamwidths than similar-sized plane-wave antennas such as parabolic dishes.

At the flare angle, the radiation of the beam lobe is down about 20 dB from its maximum value.

As the size of a horn (expressed in wavelengths) is increased, the phase error increases, giving the horn a wider radiation pattern. Keeping the beamwidth narrow requires a longer horn (smaller flare angle) to keep the phase error constant. The increasing phase error limits the aperture size of practical horns to about 15 wavelengths; larger apertures would require impractically long horns. This limits the gain of practical horns to about 1000 (30 dBi) and the corresponding minimum beamwidth to about 5 - 10°.

Optimum Horn

For a given frequency and horn length, there is some flare angle that gives minimum reflection and maximum gain. The internal reflections in straight-sided horns come from the two locations along the wave path where the impedance changes abruptly; the mouth or aperture of the horn, and the throat where the sides begin to flare out. The amount of reflection at these two sites varies with the *flare angle* of the horn (the angle the sides make with the axis). In narrow horns with small flare angles most of the reflection occurs at the mouth of the horn. The gain of the antenna is low because the small mouth approximates an open-ended waveguide. As the angle is increased, the reflection at the mouth decreases rapidly and the antenna's gain increases. In contrast, in wide horns with flare angles approaching 90° most of the reflection is at the throat. The horn's gain is again low because the throat approximates an open-ended waveguide. As the angle is decreased, the amount of reflection at this site drops, and the horn's gain again increases.

Corrugated horn antenna with a bandwidth of 3.7 to 6 GHz designed to attach to SMA
waveguide feedline. This was used as a feedhorn for a parabolic antenna on a British military base.

There is some flare angle between 0° and 90° which gives maximum gain and minimum reflection.
This is called the *optimum horn*. Most practical horn antennas are designed as optimum horns. In
a pyramidal horn, the dimensions that give an optimum horn are:

$$a_E = \sqrt{2\lambda L_E} \qquad a_H = \sqrt{3\lambda L_H}$$

For a conical horn, the dimensions that give an optimum horn are:

$$d = \sqrt{3\lambda L}$$

where:

- a_E is the width of the aperture in the E-field direction.

- a_H is the width of the aperture in the H-field direction.

- L_E is the slant length of the side in the E-field direction.

- L_H is the slant length of the side in the H-field direction.

- d is the diameter of the cylindrical horn aperture.

- L is the slant length of the cone from the apex.

- λ is the wavelength.

An optimum horn does not yield maximum gain for a given *aperture size*. That is achieved with
a very long horn (an *aperture limited* horn). The optimum horn yields maximum gain for a given
horn *length*.

Gain

Horns have very little loss, so the directivity of a horn is roughly equal to its gain. The gain G of a
pyramidal horn antenna (the ratio of the radiated power intensity along its beam axis to the intensity of an isotropic antenna with the same input power) is:

$$G = \frac{4\pi A}{\lambda^2} e_A$$

For conical horns, the gain is:

$$G = \left(\frac{\pi d}{\lambda}\right)^2 e_A$$

where:

- A is the area of the aperture,

- d is the aperture diameter of a conical horn

- λ is the wavelength,

- e_A is a dimensionless parameter between 0 and 1 called the *aperture efficiency,*

The aperture efficiency ranges from 0.4 to 0.8 in practical horn antennas. For optimum pyramidal horns, e_A = 0.511., while for optimum conical horns e_A = 0.522. So an approximate figure of 0.5 is often used. The aperture efficiency increases with the length of the horn, and for aperture-limited horns is approximately unity.

Horn-reflector Antenna

A type of antenna that combines a horn with a parabolic reflector is known as a Hogg-horn, or horn-reflector antenna, invented by Alfred C. Beck and Harald T. Friis in 1941 and further developed by David C. Hogg at Bell labs in 1961. It is also referred to as the "sugar scoop" due to its characteristic shape. It consists of a horn antenna with a reflector mounted in the mouth of the horn at a 45 degree angle so the radiated beam is at right angles to the horn axis. The reflector is a segment of a parabolic reflector, and the focus of the reflector is at the apex of the horn, so the device is equivalent to a parabolic antenna fed off-axis. The advantage of this design over a standard parabolic antenna is that the horn shields the antenna from radiation coming from angles outside the main beam axis, so its radiation pattern has very small sidelobes. Also, the aperture isn't partially obstructed by the feed and its supports, as with ordinary front-fed parabolic dishes, allowing it to achieve aperture efficiencies of 70% as opposed to 55-60% for front-fed dishes. The disadvantage is that it is far larger and heavier for a given aperture area than a parabolic dish, and must be mounted on a cumbersome turntable to be fully steerable. This design was used for a few radio telescopes and communication satellite ground antennas during the 1960s. Its largest use, however, was as fixed antennas for microwave relay links in the AT&T Long Lines microwave network. Since the 1970s this design has been superseded by shrouded parabolic dish antennas, which can achieve equally good sidelobe performance with a lighter more compact construction. Probably the most photographed and well-known example is the 15 meter (50 foot) long Holmdel Horn Antenna at Bell Labs in Holmdel, New Jersey, with which Arno Penzias and Robert Wilson discovered cosmic microwave background radiation in 1965, for which they won the 1978 Nobel Prize in Physics. Another more recent horn-reflector design is the cass-horn, which is a combination of a horn with a cassegrain parabolic antenna using two reflectors.

References

- L17-Aperture, current-lectures, antenna-dload, nikolova, faculty: ece.mcmaster.ca, Retrieved 16, June 2019

- Stutzman, Warren L.; Gary A. Thiele (2012). Antenna Theory and Design, 3rd Ed. US: John Wiley & Sons. Pp. 391–392. ISBN 978-0470576649

- Slot-Antenna, electronic-engineering, technical-references: idc-online.com , Retrieved 23, January 2019

- Kraus, John Daniel; Marhefka, Ronald J. (2002). Antennas for all applications. Mcgraw-Hill. ISBN 9780072321036

- KS-15676 Horn-Reflector Antenna Description" (PDF). Bell System Practices, Issue 3, Section 402-421-100. AT&T Co. September 1975. Retrieved 2011-12-20. On Albert lafrance [long-lines.net] website

- Poole, Ian. "Horn antenna". Radio-Electronics.com website. Adrio Communications Ltd. Retrieved 2010-11-11

5

Antenna Arrays

A collective network of multiple antennas which works together as a single unit for propagation of radio waves is termed as an antenna array. A few of its types include collinear array, parasitic array, phased array, curtain array, end-fire array, etc. All these types of antenna arrays have been carefully analyzed in this chapter.

An antenna array (often called a 'phased array') is a set of 2 or more antennas. The signals from the antennas are combined or processed in order to achieve improved performance over that of a single antenna. The antenna array can be used to:

- Increase the overall gain.

- Provide diversity reception.

- Cancel out interference from a particular set of directions.

- "Steer" the array so that it is most sensitive in a particular direction.

- Determine the direction of arrival of the incoming signals.

- To maximize the signal to Interference Plus Noise Ratio (SINR).

An antenna array is a set of N spatially separated antennas. The number of antennas in an array can be as small as 2, or as large as several thousand (as in the AN/FPS-85 Phased Array Radar Facility operated by U. S. Air Force). In general, the performance of an antenna array (for whatever application it is being used) increases with the number of antennas (elements) in the array; the drawback of course is the increased cost, size, and complexity.

The following figures show some examples of antenna arrays.

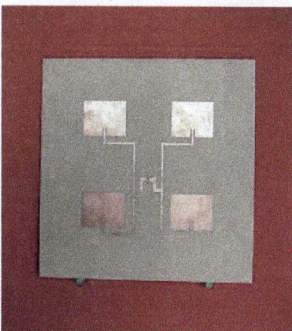

Four-element microstrip antenna array (phased array).

Cell-tower Antenna Array. These Antenna Arrays are typically used in groups of 3 (2 receive antennas and 1 transmit antenna).

The general form of an antenna array can be illustrated as in figure. An origin and coordinate system are selected, and then the N elements are positioned, each at location given by:

$$d_n = [x_n y_n z_n]$$

The positions of the elements in the phased array are illustrated in the following figure.

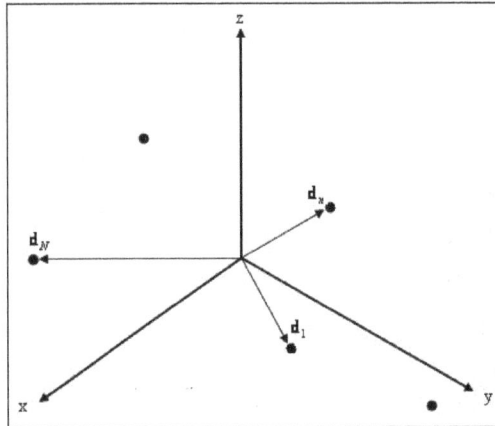

Geometry of an arbitrary N element antenna array.

Let $X_1, X_2, ..., X_N$ represent the output from antennas 1 thru N, respectively. The output from these antennas are most often multiplied by a set of N weights - $w_1, w_2, ..., w_N$ - and added together as shown in figure.

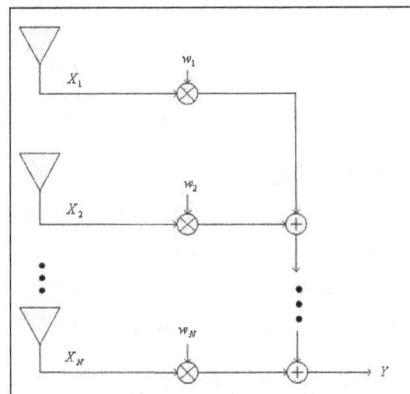

Weighting and summing of signals from the antennas to form the output in a Phased Array.

The output of an antenna array can be written succinctly as:

$$Y = \sum_{n=1}^{N} w_n X_n$$

Benefits of Antenna Arrays

To understand the benefits of antenna arrays, we will consider a set of 3-antennas located along the z-axis, receiving a signal (plane wave or the desired information) arriving from an angle relative to the z-axis θ of polar angle, as shown in figure.

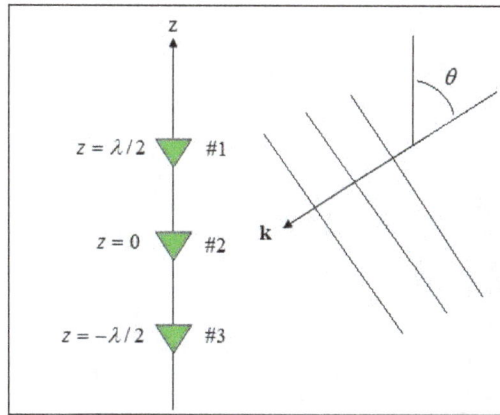

Example 3-element Antenna Array receiving a plane wave.

The antennas in the phased array are spaced one-half wavelength apart (centered at z=0). The E-field of the plane wave (assumed to have a constant amplitude everywhere) can be written as:

$$E(x,y,z) = e^{-j(k_x x + k_y y + k_z z)}$$

$$= e^{-j|k|(\sin\theta\cos\phi x + \sin\theta + \sin\phi y + \cos\theta z)}$$

$$= e^{-j k.r}$$

In the above, k is the wave vector, which specifies the variation of the phase as a function of position.

The (x,y) coordinates of each antenna is (0,0); only the z-coordinate changes for each antenna. Further, assuming that the antennas are isotropic sensors, the signal received from each antenna is proportional to the E-field at the antenna location. Hence, for antenna i, the received signal is:

$$X_i = e^{-j\frac{2\pi}{\lambda}\cos\theta z_i}$$

The received signals are distinct by a complex phase factor, which depends on the antenna separations and the angle of arrival on the plane wave. If the signals are summed together, the result is:

$$Y = e^{-j\frac{2\pi}{\lambda}\cos\theta z_1} + e^{-j\frac{2\pi}{\lambda}\cos\theta z_2} + e^{-j\frac{2\pi}{\lambda}\cos\theta z_3}$$

$$= \sum_{i=1}^{3} e^{-j\frac{2\pi}{\lambda}\cos\theta z_1}$$

$$= \sum_{m=-1}^{1} e^{-jm\pi\cos\theta z_1}$$

The interesting thing is if the magnitude of Y is plotted versus θ (the angle of arrival of the plane wave). The result is given in figure.

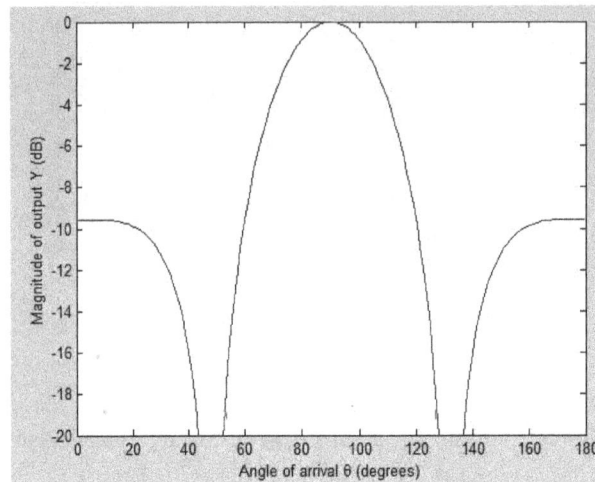

Magnitude of the output as a function of the arrival angle for Antenna Array.

In figure shows that the phased array actually processes the signals better in some directions than others. For instance, the antenna array is most receptive when the angle of arrival is 90 degrees. In contrast, when the angle of arrival is 45 or 135 degrees, the antenna array has zero output power, no matter how much power is in the incident plane wave. In this manner, a directional radiation pattern is obtained even though the antennas were assumed to be isotropic. Even though this was shown for receiving antennas, due to reciprocity, the transmitting properties would be the same.

The value and utility of an antenna array lies in its ability to determine (or alter) the received or transmitted power as a function of the arrival angle.

By choosing the weights and geometry of an antenna array properly, the phased array can be designed to cancel out energy from undesirable directions and receive energy most sensitively from other directions.

COLLINEAR ARRAY

In telecommunications, a collinear antenna array is an array of dipole antennas mounted in such a manner that the corresponding elements of each antenna are parallel and collinear, that is they are located along a common line or axis.

Collinear arrays of dipoles are high gain omnidirectional antennas. A dipole has an omnidirectional radiation pattern in free space when oriented vertically; it radiates equal radio power in all azimuthal directions perpendicular to the antenna, with the signal strength dropping to zero on the antenna axis. The purpose of stacking multiple dipoles in a vertical collinear array is to increase the power radiated in horizontal directions and reduce the power radiated into the sky or down toward the earth, where it is wasted. They radiate vertically polarized radio waves. Theoretically, when stacking idealised lossless dipole antennas in such a fashion, doubling their number will produce double the gain, with an increase of 3.01 dB. In practice, the gain realized will be below this due to imperfect radiation spread and losses.

(left & center) Collinear folded dipole arrays. Often used as base station antenna for dispatcher for police, fire, ambulance, and taxi services. *(right)* Directional antenna consisting of 4 collinear Yagi beam antennas.

Collinear dipole arrays are often used as the antennas for base stations for land mobile radio systems that communicate with mobile two-way radios in vehicles, such as police, fire, ambulance, and taxi dispatchers. They are also sometimes used for broadcasting.

Collinear dipole array on repeater for radio station JOHG-FM.

Multiple directional antennas mounted vertically separated are referred to as "stacked" and if alongside each other as "bayed".

REFLECTIVE ARRAY ANTENNA

In telecommunications and radar, a reflective array antenna is a class of directive antennas in which multiple driven elements are mounted in front of a flat surface designed to reflect the radio waves in a desired direction. They are a type of array antenna. They are often used in the VHF and UHF frequency bands. VHF examples are generally large and resemble a highway billboard, so they are sometimes called billboard antennas, or in Britain hoarding antennas. Other names are bedspring array and bowtie array depending on the type of elements making up the antenna. The curtain array is a larger version used by shortwave radio broadcasting stations.

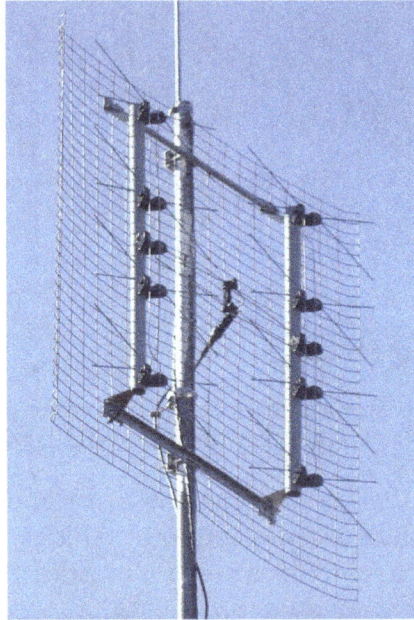

This reflective array television antenna consists of eight "bowtie" dipole driven elements mounted in front of a wire screen reflector.

The X-shaped dipoles give it a wide bandwidth to cover both the VHF (174–216 MHz) and UHF (470–700 MHz) bands. It has a gain of 5 dB VHF and 12 dB UHF and an 18 dB front-to-back ratio. The example shown is horizontally polarised.

Reflective array antennas usually have a number of identical driven elements, fed in phase, in front of a flat, electrically large reflecting surface to produce a unidirectional beam of radio waves, increasing antenna gain and reducing radiation in unwanted directions. The larger the number of elements used, the higher the gain; the narrower the beam is and the smaller the sidelobes are. The individual elements are most commonly half wave dipoles, although they sometimes contain parasitic elements as well as driven elements. The reflector may be a metal sheet or more commonly a wire screen. A metal screen reflects radio waves as well as a solid metal sheet as long as the holes in the screen are smaller than about one-tenth of a wavelength, so screens are often used to reduce weight and wind loads on the antenna. They usually consist of a grill of parallel wires or rods, oriented parallel to the axis of the dipole elements.

The driven elements are fed by a network of transmission lines, which divide the power from the RF source equally between the elements. This often has the circuit geometry of a tree structure.

Radio Signals

When a radio signal passes a conductor, it induces an electrical current in it. Since the radio signal fills space, and the conductor has a finite size, the induced currents add up or cancel out as they move along the conductor. A basic goal of antenna design is to make the currents add up to a maximum at the point where the energy is tapped off. To do this, the antenna elements are sized in relation to the wavelength of the radio signal, with the aim of setting up standing waves of current that are maximized at the feed point.

This means that an antenna designed to receive a particular wavelength has a natural size. To improve reception, one cannot simply make the antenna larger; this will improve the amount of signal intercepted by the antenna, which is largely a function of area, but will lower the efficiency of the reception (at a given wavelength). Thus, in order to improve reception, antenna designers often use multiple elements, combining them together so their signals add up. These are known as *antenna arrays*.

Array Phasing

In order for the signals to add together, they need to arrive in-phase. Consider two dipole antennas placed in a line end-to-end, or *collinear*. If the resulting array is pointed directly at the source signal, both dipoles will see the same instantaneous signal, and thus their reception will be in-phase. However, if one were to rotate the antenna so it was at an angle to the signal, the extra path from the signal to the more distant dipole means it receives the signal slightly out of phase. When the two signals are then added up, they no longer strictly reinforce each other, and the output drops. This makes the array more sensitive horizontally, while stacking the dipoles in parallel narrows the pattern vertically. This allows the designer to tailor the reception pattern, and thus the gain, by moving the elements about.

If the antenna is properly aligned with the signal, at any given instant in time, all of the elements in an array will receive the same signal and be in-phase. However, the output from each element has to be gathered up at a single feed point, and as the signals travel across the antenna to that point, their phase is changing. In a two-element array this is not a problem because the feed point can be placed between them; any phase shift taking place in the transmission lines is equal for both elements. However, if one extends this to a four-element array, this approach no longer works, as the signal from the outer pair has to travel further and will thus be at a different phase than the inner pair when it reaches the center. To ensure that they all arrive with the same phase, it is common to see additional transmission wire inserted in the signal path, or for the transmission line to be crossed over to reverse the phase if the difference is greater than 1/2 a wavelength.

Reflectors

The gain can be further improved through the addition of a *reflector*. Generally any conductor in a flat sheet will act in a mirror-like fashion for radio signals, but this also holds true for non-continuous surfaces as long as the gaps between the conductors are less than about 1/10 of the target wavelength. This means that wire mesh or even parallel wires or metal bars can be used, which is especially useful both for reducing the total amount of material as well as reducing wind loads.

Due to the change in signal propagation direction on reflection, the signal undergoes a reversal of phase. In order for the reflector to add to the output signal, it has to reach the elements in-phase. Generally this would require the reflector to be placed at 1/2 of a wavelength behind the elements, and this can be seen in many common reflector arrays like television antennas.

However, there are a number of factors that can change this distance, and actual reflector positioning varies.

Reflectors also have the advantage of reducing the signal received from the back of the antenna. Signals received from the rear and re-broadcast from the reflector have not undergone a change of phase, and do not add to the signal from the front. This greatly improves the front-to-back ratio of the antenna, making it more directional. This can be useful when a more directional signal is desired, or unwanted signals are present. There are cases when this is not desirable, and although reflectors are commonly seen in array antennas, they are not universal. For instance, while UHF television antennas often use an array of bowtie antennas with a reflector, a bowtie array without a reflector is a relatively common design in the microwave region.

Gain Limits

As more elements are added to an array, the beamwidth of the antenna's main lobe decreases, leading to an increase in gain. In theory there is no limit to this process. However, as the number of elements increases, the complexity of the required feed network that keeps the signals in-phase increases. Ultimately, the rising inherent losses in the feed network become greater than the additional gain achieved with more elements, limiting the maximum gain that can be achieved.

The gain of practical array antennas is limited to about 25 - 30 dB. Two half wave elements spaced a half wave apart and a quarter wave from a reflecting screen have been used as a standard gain antenna with about 9.8 dBi at its design frequency. Common 4-bay television antennas have gains around 10 to 12 dB, and 8-bay designs might increase this to 12 to 16 dB. The 32-element SCR-270 had a gain around 19.8 dB. Some very large reflective arrays have been constructed, notably the Soviet Duga radars which are hundreds of meters across and contain hundreds of elements. *Active* array antennas, in which groups of elements are driven by separate RF amplifiers, can have much higher gain, but are prohibitively expensive.

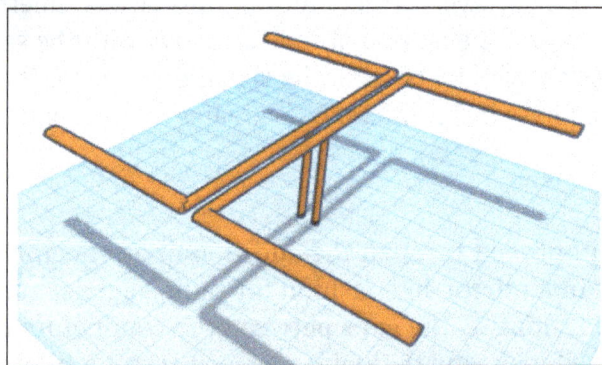

Two element dipole array in front of a one wavelength square reflector used as gain standard

Since the 1980s, versions for use at microwave frequencies have been made with patch antenna elements mounted in front of a metal surface.

Radiation Pattern and Beam Steering

When driven in phase, the radiation pattern of the reflective array is a single main lobe perpendicular to the plane of the antenna, plus several sidelobes at equal angles to either side. The more elements used, the narrower the main lobe and the less power is radiated in the sidelobes.

The main lobe of the antenna can be steered electronically within a limited angle by phase shifting the drive signals applied to the individual elements. Each antenna element is fed through a phase shifter which can be controlled digitally, delaying each signal by a successive amount. This causes the wavefronts created by the superposition of the individual elements to be at an angle to the plane of the antenna. Antennas that use this technique are called phased arrays and often used in modern radar systems.

Another option for steering the beam is mounting the entire antenna structure on a pivoting platform and rotating it mechanically.

PARASITIC ARRAY

A parasitic element is an element, which depends on other's feed. It does not have its own feed. Hence, in this type of arrays we employ such elements, which help in increasing the radiation indirectly.

These parasitic elements are not directly connected to the feed.

Construction and Working of Parasitic Array

Let us look at the important parts of a Parasitic array and how they work.

The main parts are:

- Driven element
- Parasitic elements
  - Reflector
  - Director
- Boom

Driven Element

The antennas radiate individually and while in array, the radiation of all the elements sum up to form the radiation beam. All the elements of the array need not be connected to the feed. The dipole that is connected to the feed is known as a driven element.

Parasitic Elements

The elements, which are added do not possess an electrical connection between them to the driven element or the feed. They are positioned so that they lie in the induction field of the driven element. Hence, they are known as parasitic elements.

Reflector

If one of the parasitic element, which is 5% longer than driven element, is placed close to the driven element is longer, then it acts as a concave mirror, which reflects the energy in the direction of the radiation pattern rather than its own direction and hence is known as a reflector.

Director

A parasitic element, which is 5% shorter than the driven element, from which it receives energy, tends to increase radiation in its own direction and therefore, behaves like convergent convex lens. This element is called as a director. A number of directors are placed to increase the directivity.

Boom

The element on which all these are placed is callled a boom. It is a non-metallic structure which provides insulation, so that there will not be any short circuit between the other elements of the array.

These are all the main elements, which contribute the radiation. This can be better understood with the help of a diagram.

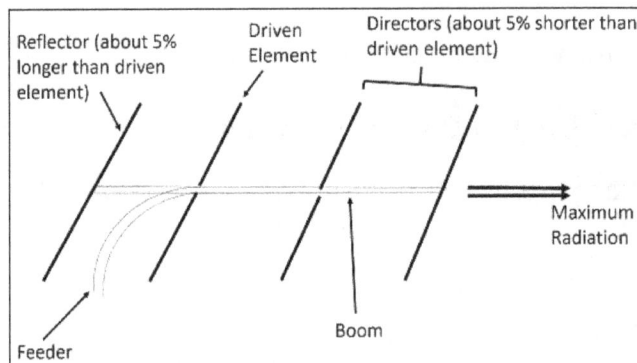

The image shown above is that of a parasitic array, which shows the parts of parsitic array such as the driven element, the directors and the reflector. The feed is given through the feeder.

The arrays are used at frequencies ranging from 2MHz to several GHz. These are especially used to get high directivity, and better forward gain with a uni-directional. The most common example of this type of array is the Yagi-Uda antenna. Quad antenna may also be quoted as another example.

PHASED ARRAY

In antenna theory, a phased array usually means an electronically scanned array, a computer-controlled array of antennas which creates a beam of radio waves that can be electronically steered to point in different directions without moving the antennas. In an array antenna, the radio frequency current from the transmitter is fed to the individual antennas with the correct phase relationship so that the radio waves from the separate antennas add together to increase the radiation in a desired direction, while cancelling to suppress radiation in undesired directions. In a phased array, the power from the transmitter is fed to the antennas through devices called *phase shifters*, controlled by a computer system, which can alter the phase electronically, thus steering the beam of radio waves to a different direction. Since the array must consist of many small antennas (sometimes thousands) to achieve high gain, phased arrays are mainly practical at the high frequency end of the radio spectrum, in the UHF and microwave bands, in which the antenna elements are conveniently small.

Phased arrays were invented for use in military radar systems, to steer a narrow antenna pattern quickly across the sky to detect planes and missiles. These phased array radar systems are now widely used, and phased arrays are spreading to civilian applications. The phased array principle is also used in acoustics, and phased arrays of acoustic transducers are used in medical ultrasound imaging scanners (phased array ultrasonics), oil and gas prospecting (reflection seismology), and military sonar systems.

The term "phased array" is also used to a lesser extent for unsteered array antennas in which the phase of the feed power and thus the radiation pattern of the antenna array is fixed. For example, AM broadcast radio antennas consisting of multiple mast radiators fed so as to create a specific radiation pattern are also called "phased arrays".

Types

Phased arrays take multiple forms. However, the four most common are the passive phased array (PESA), active electronically scanned array (AESA), hybrid beam forming phased array, and digital beam forming (DBF) array.

A *passive phased array* or *passive electronically scanned array* (PESA) is a phased array in which the antenna elements are connected to a single transmitter and/or receiver. PESAs are the most common type of phased array. Generally speaking, a PESA uses one receiver/exciter for the entire array.

An *active phased array* or *active electronically scanned array* (AESA) is a phased array in which each antenna element has an analog transmitter/receiver (T/R) module which creates the phase shifting required to electronically steer the antenna beam. Active arrays are a more advanced, second-generation phased-array technology which are used in military applications; unlike PESAs they can radiate several beams of radio waves at multiple frequencies in different directions simultaneously. However, the number of simultaneous beams is limited by practical reasons of electronic packaging of the beam former(s) to approximately three simultaneous beams for an AESA. Each beam former has a receiver/exciter connected to it.

A *hybrid beam forming phased array* can be thought of as a combination of an AESA and a digital beam forming phased array. It uses subarrays that are active phased arrays (for instance, a subarry may be 64, 128 or 256 elements and the number of elements depends upon system requirements). The subarrays are combined together to form the full array. Each subarray has its own digital receiver/exciter. This approach allows clusters of simultaneous beams to be created.

A *digital beam forming (DBF) phased array* has a digital receiver/exciter at each element in the array. The signal at each element is digitized by the receiver/exciter. This means that antenna beams can be formed digitally in a field programmable gate array (FPGA) or the array computer. This approach allows for multiple simultaneous antenna beams to be formed.

One possible physical implementation of a phased array is called a *conformal antenna*. It is a phased array in which the individual antennas, instead of being arranged in a flat plane, are mounted on a curved surface. The phase shifters compensate for the different path lengths of the waves due to the antenna elements' varying position on the surface, allowing the array to radiate a plane wave. Conformal antennas are used in aircraft and missiles, to integrate the antenna into the curving surface of the aircraft to reduce aerodynamic drag.

Applications

Broadcasting

In broadcast engineering, phased arrays are used by many AM broadcast radio stations to enhance signal strength and therefore coverage in the city of license, while minimizing interference to other areas. Due to the differences between daytime and nighttime ionospheric propagation at medium-wave frequencies, it is common for AM broadcast stations to change between day (groundwave) and night (skywave) radiation patterns by switching the phase and power levels supplied to the individual antenna elements (mast radiators) daily at sunrise and sunset. For shortwave broadcasts many stations use arrays of horizontal dipoles. A common arrangement uses 16 dipoles in a 4×4 array. Usually this is in front of a wire grid reflector. The phasing is often switchable to allow Beam steering in azimuth and sometimes elevation.

More modest phased array longwire antenna systems may be employed by private radio enthusiasts to receive longwave, mediumwave (AM) and shortwave radio broadcasts from great distances.

On VHF, phased arrays are used extensively for FM broadcasting. These greatly increase the antenna gain, magnifying the emitted RF energy toward the horizon, which in turn greatly increases a station's broadcast range. In these situations, the distance to each element from the transmitter is identical, or is one (or other integer) wavelength apart. Phasing the array such that the lower elements are slightly delayed (by making the distance to them longer) causes a downward beam tilt, which is very useful if the antenna is quite high on a radio tower.

Other phasing adjustments can increase the downward radiation in the far field without tilting the main lobe, creating null fill to compensate for extremely high mountaintop locations, or decrease it in the near field, to prevent excessive exposure to those workers or even nearby homeowners on the ground. The latter effect is also achieved by half-wave spacing – inserting additional elements halfway between existing elements with full-wave spacing. This phasing achieves roughly the same horizontal gain as the full-wave spacing; that is, a five-element full-wave-spaced array equals a nine- or ten-element half-wave-spaced array.

Radar

Phased array radar systems are also used by warships of many navies. Because of the rapidity with which the beam can be steered, phased array radars allow a warship to use one radar system for surface detection and tracking (finding ships), air detection and tracking (finding aircraft and missiles) and missile uplink capabilities. Before using these systems, each surface-to-air missile in flight required a dedicated fire-control radar, which meant that radar-guided weapons could only engage a small number of simultaneous targets. Phased array systems can be used to control missiles during the mid-course phase of the missile's flight. During the terminal portion of the flight, continuous-wave fire control directors provide the final guidance to the target. Because the antenna pattern is electronically steered, phased array systems can direct radar beams fast enough to maintain a fire control quality track on many targets simultaneously while also controlling several in-flight missiles.

The AN/SPY-1 phased array radar, part of the Aegis Combat System deployed on modern U.S. cruisers and destroyers, "is able to perform search, track and missile guidance functions simultaneously with a capability of over 100 targets." Likewise, the Thales Herakles phased array multi-function radar used in service with France, Russia and Singapore has a track capacity of 200 targets and is able to achieve automatic target detection, confirmation and track initiation in a single scan, while simultaneously providing mid-course guidance updates to the MBDA Aster missiles launched from the ship. The German Navy and the Royal Dutch Navy have developed the Active Phased Array Radar System (APAR). The MIM-104 Patriot and other ground-based antiaircraft systems use phased array radar for similar benefits.

Active Phased Array Radar mounted on top of *Sachsen*-class frigate F220 *Hamburg's* superstructure of the German Navy.

Phased arrays are used in naval sonar, in active (transmit and receive) and passive (receive only) and hull-mounted and towed array sonar.

Space Probe Communication

The *Messenger* spacecraft was a space probe mission to the planet Mercury. This was the first deep-space mission to use a phased-array antenna for communications. The radiating elements are circularly-polarized, slotted waveguides. The antenna, which uses the X band, used 26 radiative elements and can gracefully degrade.

Weather Research Usage

The National Severe Storms Laboratory has been using a SPY-1A phased array antenna, provided by the US Navy, for weather research at its Norman, Oklahoma facility since April 23, 2003. It is hoped that research will lead to a better understanding of thunderstorms and tornadoes, eventually leading to increased warning times and enhanced prediction of tornadoes. Current project participants include the National Severe Storms Laboratory and National Weather Service

Radar Operations Center, Lockheed Martin, United States Navy, University of Oklahoma School of Meteorology, School of Electrical and Computer Engineering, and Atmospheric Radar Research Center, Oklahoma State Regents for Higher Education, the Federal Aviation Administration, and Basic Commerce and Industries. The project includes research and development, future technology transfer and potential deployment of the system throughout the United States. It is expected to take 10 to 15 years to complete and initial construction was approximately $25 million. A team from Japan's RIKEN Advanced Institute for Computational Science (AICS) has begun experimental work on using phased-array radar with a new algorithm for instant weather forecasts.

AN/SPY-1A radar installation at National Severe Storms Laboratory, Norman, Oklahoma. The enclosing radome provides weather protection.

Optics

Within the visible or infrared spectrum of electromagnetic waves it is possible to construct optical phased arrays. They are used in wavelength multiplexers and filters for telecommunication purposes, laser beam steering, and holography. Synthetic array heterodyne detection is an efficient method for multiplexing an entire phased array onto a single element photodetector. The dynamic beam forming in an optical phased array transmitter can be used to electronically raster or vector scan images without using lenses or mechanically moving parts in a lensless projector. Optical phased array receivers have been demonstrated to be able to act as lensless cameras by selectively looking at different directions.

Satellite Broadband Internet Transceivers

OneWeb and Starlink are two low-earth orbit satellite constellations which are under construction as of 2019. They are designed to provide broadband internet connectivity to consumers; the user terminals of both systems will use phased array antennas.

Radio-frequency Identification

By 2014, phased array antennas were integrated into RFID systems to increase the area of coverage of a single system by 100% to 76,200 m² (820,000 sq ft) while still using traditional passive UHF tags.

Human-machine Interfaces

A phased array of acoustic transducers, denominated airborne ultrasound tactile display (AUTD), was developed in 2008 at the University of Tokyo's Shinoda Lab to induce tactile feedback. This system was demonstrated to enable a user to interactively manipulate virtual holographic objects.

Mathematical Perspective and Formulas

Mathematically a phased array is an example of N-slit diffraction, in which the radiation field at the receiving point is the result of the coherent addition of N point sources in a line. Since each individual antenna acts as a slit, emitting radio waves, their diffraction pattern can be calculated by adding the phase shift φ to the fringing term.

We will begin from the N-slit diffraction pattern derived on the diffraction formalism page, with N slits of equal size a and spacing d.

$$\psi = \psi_0 \frac{\sin\left(\frac{\pi a}{\lambda}\sin\theta\right)}{\frac{\pi a}{\lambda}\sin\theta} \frac{\sin\left(\frac{N}{2}kd\sin\theta\right)}{\sin\left(\frac{kd}{2}\sin\theta\right)}$$

Now, adding a φ term to the $kd\sin\theta$ fringe effect in the second term yields:

$$\psi = \psi_0 \frac{\sin\left(\frac{\pi a}{\lambda}\sin\theta\right)}{\frac{\pi a}{\lambda}\sin\theta} \frac{\sin\left(\frac{N}{2}\left(\frac{2\pi d}{\lambda}\sin\theta + \phi\right)\right)}{\sin\left(\frac{\pi d}{\lambda}\sin\theta + \frac{\phi}{2}\right)}$$

Taking the square of the wave function gives us the intensity of the wave.

$$I = I_0 \left(\frac{\sin\left(\frac{\pi a}{\lambda}\sin\theta\right)}{\frac{\pi a}{\lambda}\sin\theta}\right)^2 \left(\frac{\sin\left(\frac{N}{2}\left(\frac{2\pi d}{\lambda}\sin\theta + \phi\right)\right)}{\sin\left(\frac{\pi d}{\lambda}\sin\theta + \frac{\phi}{2}\right)}\right)^2$$

$$I = I_0 \left(\frac{\sin\left(\frac{\pi a}{\lambda}\sin\theta\right)}{\frac{\pi a}{\lambda}\sin\theta}\right)^2 \left(\frac{\sin\left(\frac{\pi}{\lambda}Nd\sin\theta + \frac{N}{2}\phi\right)}{\sin\left(\frac{\pi d}{\lambda}\sin\theta + \frac{\phi}{2}\right)}\right)^2$$

Now space the emitters a distance $d = \dfrac{\lambda}{4}$ apart. This distance is chosen for simplicity of calculation but can be adjusted as any scalar fraction of the wavelength.

$$I = I_0 \left(\frac{\sin\left(\dfrac{\pi a}{\lambda} \sin\theta \right)}{\dfrac{\pi a}{\lambda} \sin\theta} \right)^2 \left(\frac{\sin\left(\dfrac{\pi}{4} N \sin\theta + \dfrac{N}{2}\phi \right)}{\sin\left(\dfrac{\pi}{4} \sin\theta + \dfrac{\phi}{2} \right)} \right)^2$$

As sine achieves its maximum at $\dfrac{\pi}{2}$, we set the numerator of the second term = 1.

$$\frac{\pi}{4} N \sin\theta + \frac{N}{2}\phi = \frac{\pi}{2}$$

$$\sin\theta = \left(\frac{\pi}{2} - \frac{N}{2}\phi \right) \frac{4}{N\pi}$$

$$\sin\theta = \frac{2}{N} - \frac{2\phi}{\pi}$$

Thus as N gets large, the term will be dominated by the $\dfrac{2\phi}{\pi}$ term. As sine can oscillate between -1 and 1, we can see that setting $\phi = -\dfrac{\pi}{2}$ will send the maximum energy on an angle given by,

$$\theta = \sin^{-1} 1 = \frac{\pi}{2} = 90°$$

Additionally, we can see that if we wish to adjust the angle at which the maximum energy is emitted, we need only to adjust the phase shift φ between successive antennas. Indeed, the phase shift corresponds to the negative angle of maximum signal.

A similar calculation will show that the denominator is minimized by the same factor.

Different Types of Phased Arrays

There are two main types of beamformers. These are time domain beamformers and frequency domain beamformers.

A graduated attenuation window is sometimes applied across the face of the array to improve side-lobe suppression performance, in addition to the phase shift.

Time domain beamformer works by introducing time delays. The basic operation is called "delay and sum". It delays the incoming signal from each array element by a certain amount of time, and then adds them together. A Butler matrix allows several beams to be formed simultaneously, or one beam to be scanned through an arc. The most common kind of time domain beam former is serpentine waveguide. Active phased array designs use individual delay lines that are switched on and off. Yttrium iron garnet phase shifters vary the phase delay using the strength of a magnetic field.

There are two different types of frequency domain beamformers:

- The first type separates the different frequency components that are present in the received signal into multiple frequency bins (using either a Discrete Fourier transform (DFT) or a filterbank). When different delay and sum beamformers are applied to each frequency bin, the result is that the main lobe simultaneously points in multiple different directions at each of the different frequencies. This can be an advantage for communication links, and is used with the SPS-48 radar.

- The other type of frequency domain beamformer makes use of Spatial Frequency. Discrete samples are taken from each of the individual array elements. The samples are processed using a DFT. The DFT introduces multiple different discrete phase shifts during processing. The outputs of the DFT are individual channels that correspond with evenly spaced beams formed simultaneously. A 1-dimensional DFT produces a fan of different beams. A 2-dimensional DFT produces beams with a pineapple configuration.

These techniques are used to create two kinds of phased array:

- Dynamic – an array of variable phase shifters are used to move the beam.

- Fixed – the beam position is stationary with respect to the array face and the whole antenna is moved.

There are two further sub-categories that modify the kind of dynamic array or fixed array:

- Active – amplifiers or processors are in each phase shifter element.

- Passive – large central amplifier with attenuating phase shifters.

Dynamic Phased Array

Each array element incorporates an adjustable phase shifter that are collectively used to move the beam with respect to the array face.

Dynamic phased array require no physical movement to aim the beam. The beam is moved electronically. This can produce antenna motion fast enough to use a small pencil-beam to simultaneously track multiple targets while searching for new targets using just one radar set (track while search).

As an example, an antenna with a 2 degree beam with a pulse rate of 1 kHz will require approximately 8 seconds to cover an entire hemisphere consisting of 8,000 pointing positions. This configuration provides 12 opportunities to detect a 1,000 m/s (2,200 mph; 3,600 km/h) vehicle over a range of 100 km (62 mi), which is suitable for military applications.

The position of mechanically steered antennas can be predicted, which can be used to create electronic countermeasures that interfere with radar operation. The flexibility resulting from phased array operation allows beams to be aimed at random locations, which eliminates this vulnerability. This is also desirable for military applications.

Fixed Phased Array

Fixed phased array antennas are typically used to create an antenna with a more desirable form factor than the conventional parabolic reflector or cassegrain reflector. Fixed phased arrays incorporate fixed phase shifters. For example, most commercial FM Radio and TV antenna towers use a collinear antenna array, which is a fixed phased array of dipole elements.

An antenna tower consisting of a fixed phase collinear antenna array with four elements.

In radar applications, this kind of phased array is physically moved during the track and scan process. There are two configurations:

- Multiple frequencies with a delay-line.

- Multiple adjacent beams.

The SPS-48 radar uses multiple transmit frequencies with a serpentine delay line along the left side of the array to produce vertical fan of stacked beams. Each frequency experiences a different phase shift as it propagates down the serpentine delay line, which forms different beams. A filter bank is used to split apart the individual receive beams. The antenna is mechanically rotated.

Semi-active radar homing uses monopulse radar that relies on a fixed phased array to produce multiple adjacent beams that measure angle errors. This form factor is suitable for gimbal mounting in missile seekers.

Active Phased Array

Active electronically-scanned arrays (AESA) elements incorporate transmit amplification with phase shift in each antenna element (or group of elements). Each element also includes receive pre-amplification. The phase shifter setting is the same for transmit and receive.

Active phased arrays do not require phase reset after the end of the transmit pulse, which is compatible with Doppler radar and pulse-Doppler radar.

Passive Phased Array

Passive phased arrays typically use large amplifiers that produce all of the microwave transmit signal for the antenna. Phase shifters typically consist of waveguide elements controlled by magnetic field, voltage gradient, or equivalent technology.

The phase shift process used with passive phased arrays typically puts the receive beam and transmit beam into diagonally opposite quadrants. The sign of the phase shift must be inverted after the transmit pulse is finished and before the receive period begins to place the receive beam into the same location as the transmit beam. That requires a phase impulse that degrades sub-clutter visibility performance on Doppler radar and Pulse-Doppler radar. As an example, Yttrium iron garnet phase shifters must be changed after transmit pulse quench and before receiver processing starts to align transmit and receive beams. That impulse introduces FM noise that degrades clutter performance.

Passive phased array design is used in the AEGIS Combat System. for direction-of-arrival estimation.

Active Electronically Scanned Array

An active electronically scanned array (AESA) is a type of phased array antenna, which is a computer-controlled array antenna in which the beam of radio waves can be electronically steered to point in different directions without moving the antenna. In the AESA, each antenna element is connected to a small solid-state transmit/receive module (TRM) under the control of a computer, which performs the functions of a transmitter and/or receiver for the antenna. This contrasts with a passive electronically scanned array (PESA), in which all the antenna elements are connected to a single transmitter and/or receiver through phase shifters under the control of the computer. AESA's main use is in radar, and these are known as active phased array radar (APAR).

The Eurofighter Typhoon combat aircraft with its nose fairing removed, revealing its Euroradar CAPTOR AESA radar antenna.

The AESA is a more advanced, sophisticated, second-generation of the original PESA phased array technology. PESAs can only emit a single beam of radio waves at a single frequency at a time. The AESA can radiate multiple beams of radio waves at multiple frequencies simultaneously. AESA radars can spread their signal emissions across a wider range of frequencies, which makes them more difficult to detect over background noise, allowing ships and aircraft to radiate powerful radar signals while still remaining stealthy.

Radar systems generally work by connecting an antenna to a powerful radio transmitter to emit a short pulse of signal. The transmitter is then disconnected and the antenna is connected to a sensitive receiver which amplifies any echos from target objects. By measuring the time it takes for the signal to return, the radar receiver can determine the distance to the object. The receiver then sends the resulting output to a display of some sort. The transmitter elements were typically klystron tubes or magnetrons, which are suitable for amplifying or generating a narrow range of frequencies to high power levels. To scan a portion of the sky, the radar antenna must be physically moved to point in different directions.

Starting in the 1960s new solid-state devices capable of delaying the transmitter signal in a controlled way were introduced. That led to the first practical large-scale passive electronically scanned array (PESA), or simply phased array radar. PESAs took a signal from a single source, split it into hundreds of paths, selectively delayed some of them, and sent them to individual antennas. The radio signals from the separate antennas overlapped in space, and the interference patterns between the individual signals was controlled to reinforce the signal in certain directions, and mute it in all others. The delays could be easily controlled electronically, allowing the beam to be steered very quickly without moving the antenna. A PESA can scan a volume of space much quicker than a traditional mechanical system. Additionally, thanks to progress in electronics, PESAs added the ability to produce several active beams, allowing them to continue scanning the sky while at the same time focusing smaller beams on certain targets for tracking or guiding semi-active radar homing missiles. PESAs quickly became widespread on ships and large fixed emplacements in the 1960s, followed by airborne sensors as the electronics shrank.

AESAs are the result of further developments in solid-state electronics. In earlier systems the transmitted signal was originally created in a klystron or traveling wave tube or similar device, which are relatively large. Receiver electronics were also large due to the high frequencies that they worked with. The introduction of gallium arsenide microelectronics through the 1980s served to greatly reduce the size of the receiver elements, until effective ones could be built at sizes similar to those of handheld radios, only a few cubic centimeters in volume. The introduction of JFETs and MESFETs did the same to the transmitter side of the systems as well. It gave rise to Amplifier-Transmitters with a low-power solid state waveform generator feeding an amplifier, allowing any radar so equipped to transmit on a much wider range of frequencies, to the point of changing operating frequency with every pulse sent out. Shrinking the entire assembly (the transmitter, receiver and antenna) into a single "transmitter-receiver module" (TRM) about the size of a carton of milk and arraying these elements produces an AESA.

The primary advantage of an AESA over a PESA is capability of the different modules to operate on different frequencies. Unlike the PESA, where the signal is generated at single frequencies by a small number of transmitters, in the AESA each module generates and radiates its own independent signal. This allows the AESA to produce numerous simultaneous "sub-beams" that it can recognize due to different frequencies, and actively track a much larger number of targets. AESAs can also produce beams that consist of many different frequencies at once, using post-processing of the combined signal from a number of TRMs to re-create a display as if there was a single powerful beam being sent. However, this means that the noise present in each frequency is also received and added.

Advantages

AESAs add many capabilities of their own to those of the PESAs. Among these are: the ability to form multiple beams simultaneously, to use groups of TRMs for different roles concurrently, like radar detection, and, more importantly, their multiple simultaneous beams and scanning frequencies create difficulties for traditional, correlation-type radar detectors.

Low Probability of Intercept

Radar systems work by sending out a signal and then listening for its echo off distant objects. Each of these paths, to and from the target, is subject to the inverse square law of propagation in both the transmitted signal and the signal reflected back. That means that a radar's received energy drops with the fourth power of the distance, which is why radar systems require high powers, often in the megawatt range, to be effective at long range.

The radar signal being sent out is a simple radio signal, and can be received with a simple radio receiver. Military aircraft and ships have defensive receivers, called "radar warning receivers" (RWR), which detect when an enemy radar beam is on them, thus revealing the position of the enemy. Unlike the radar unit, which must send the pulse out and then receive its reflection, the target's receiver does not need the reflection and thus the signal drops off only as the square of distance. This means that the receiver is always at an advantage neglecting disparity in antenna size over the radar in terms of range - it will always be able to detect the signal long before the radar can see the target's echo. Since the position of the radar is extremely useful information in an attack on that platform, this means that radars generally must be turned off for lengthy periods if they are subject to attack; this is common on ships, for instance.

Unlike the radar, which knows which direction it is sending its signal, the receiver simply gets a pulse of energy and has to interpret it. Since the radio spectrum is filled with noise, the receiver's signal is integrated over a short period of time, making periodic sources like a radar add up and stand out over the random background. The rough direction can be calculated using a rotating antenna, or similar passive array using phase or amplitude comparison. Typically RWRs store the detected pulses for a short period of time, and compare their broadcast frequency and pulse repetition frequency against a database of known radars. The direction to the source is normally combined with symbology indicating the likely purpose of the radar – Airborne early warning and control, surface-to-air missile, etc.

This technique is much less useful against a radar with a frequency-agile (solid state) transmitter. Since the AESA (or PESA) can change its frequency with every pulse (except when using doppler filtering), and generally does so using a random sequence, integrating over time does not help pull the signal out of the background noise. Moreover, a radar may be designed to extend the duration of the pulse and lower its peak power. An AESA or modern PESA will often have the capability to alter these parameters during operation. This makes no difference to the total energy reflected by the target but makes the detection of the pulse by an RWR system less likely. Nor does the AESA have any sort of fixed pulse repetition frequency, which can also be varied and thus hide any periodic brightening across the entire spectrum. Older generation RWRs are essentially useless against AESA radars, which is why AESA's are also known as 'low probability of intercept radars. Modern RWRs must be made highly sensitive (small angles and bandwidths for individual antennas, low transmission loss and noise) and add successive pulses through time-frequency processing to achieve useful detection rates.

High Jamming Resistance

Jamming is likewise much more difficult against an AESA. Traditionally, jammers have operated by determining the operating frequency of the radar and then broadcasting a signal on it to confuse the receiver as to which is the "real" pulse and which is the jammer's. This technique works as long as the radar system cannot easily change its operating frequency. When the transmitters were based on klystron tubes this was generally true, and radars, especially airborne ones, had only a few frequencies to choose among. A jammer could listen to those possible frequencies and select the one to be used to jam.

Most radars using modern electronics are capable of changing their operating frequency with every pulse. This can make jamming less effective; although it is possible to send out broadband white noise to conduct barrage jamming against all the possible frequencies, this reduces the amount of jammer energy in any one frequency. An AESA has the additional capability of spreading its frequencies across a wide band even in a single pulse, a technique known as a "chirp". In this case, the jamming will be the same frequency as the radar for only a short period, while the rest of the radar pulse is unjammed.

AESAs can also be switched to a receive-only mode, and use these powerful jamming signals to track its source, something that required a separate receiver in older platforms. By integrating received signals from the targets' own radar along with a lower rate of data from its own broadcasts, a detection system with a precise RWR like an AESA can generate more data with less energy. Some receive beamforming-capable systems, usually ground-based, may even discard a transmitter entirely.

However, using a single receiving antenna only gives a direction. Obtaining a range and a target vector requires at least two physically separate passive devices for triangulation to provide instantaneous determinations, unless phase interferometry is used. Target motion analysis can estimate these quantities by incorporating many directional measurements over time, along with knowledge of the position of the receiver and constraints on the possible motion of the target.

Other Advantages

Since each element in an AESA is a powerful radio receiver, active arrays have many roles besides traditional radar. One use is to dedicate several of the elements to reception of common radar signals, eliminating the need for a separate radar warning receiver. The same basic concept can be used to provide traditional radio support, and with some elements also broadcasting, form a very high bandwidth data link. The F-35 uses this mechanism to send sensor data between aircraft in order to provide a synthetic picture of higher resolution and range than any one radar could generate. In 2007, tests by Northrop Grumman, Lockheed Martin, and L-3 Communications enabled the AESA system of a Raptor to act like a WiFi access point, able to transmit data at 548 megabits per second and receive at gigabit speed; this is far faster than the Link 16 system used by US and allied aircraft, which transfers data at just over 1 Mbit/s. To achieve these high data rates requires a highly directional antenna which AESA provides but which precludes reception by other units not within the antennas beamwidth, whereas like most Wi-Fi designs, Link-16 transmits its signal omni-directionally to ensure all units within range can receive the data.

AESAs are also much more reliable than either a PESA or older designs. Since each module operates independently of the others, single failures have little effect on the operation of the system as a whole. Additionally, the modules individually operate at low powers, perhaps 40 to 60 watts, so the need for a large high-voltage power supply is eliminated.

Replacing a mechanically scanned array with a fixed AESA mount (such as on the Boeing F/A-18E/F Super Hornet) can help reduce an aircraft's overall radar cross-section (RCS), but some designs (such as the Eurofighter Typhoon) forgo this advantage in order to combine mechanical scanning with electronic scanning and provide a wider angle of total coverage. This high off-nose pointing allows the AESA equipped fighter to employ a Crossing the T maneuver, often referred to as 'beaming' in the context of air-to-air combat, against a mechanically scanned radar that would filter out the low closing speed of the perpendicular flight as ground clutter while the AESA swivels 40 degrees towards the target in order to keep it within the AESA's 60 degree off-angle limit.

Limitations

With a half wavelength distance between the elements, the maximum beam angle is approximately ±45°. With a shorter element distance, the highest Field of View (FOV) for a flat phased array antenna is currently 120° (±60°), although this can be combined with mechanical steering as noted above.

CONFORMAL ARRAY

In radio communication and avionics a conformal antenna or conformal array is a flat radio antenna which is designed to conform or follow some prescribed shape, for example a flat curving antenna which is mounted on or embedded in a curved surface. Conformal antennas were developed in the 1980s as avionics antennas integrated into the curving skin of military aircraft to reduce aerodynamic drag, replacing conventional antenna designs which project from the aircraft surface. Military aircraft and missiles are the largest application of conformal antennas, but they are also used in some civilian aircraft, military ships and land vehicles. As the cost of the required processing technology comes down, they are being considered for use in civilian applications such as train antennas, car radio antennas, and cellular base station antennas, to save space and also to make the antenna less visually intrusive by integrating it into existing objects.

Working

Conformal antennas are a form of phased array antenna. They are composed of an array of many identical small flat antenna elements, such as dipole, horn, or patch antennas, covering the surface. At each antenna the current from the transmitter passes through a phase shifter device which are all controlled by a microprocessor (computer). By controlling the phase of the feed current, the nondirectional radio waves emitted by the individual antennas can be made to combine in front of the antenna by the process of interference, forming a strong beam (or beams) of radio waves pointed in any desired direction. In a receiving antenna the weak individual radio signals received by each antenna element are combined in the correct phase to enhance signals coming from a particular direction, so the antenna can be made sensitive to the signal from a particular station and reject interfering signals from other directions.

In a conventional phased array the individual antenna elements are mounted on a flat surface. In a conformal antenna, they are mounted on a curved surface, and the phase shifters also compensate for the different phase shifts caused by the varying path lengths of the radio waves due to the location of the individual antennas on the curved surface. Because the individual antenna elements must be small, conformal arrays are typically limited to high frequencies in the UHF or microwave range, where the wavelength of the waves is small enough that small antennas can be used.

LOG-PERIODIC ARRAY

A log-periodic antenna (LP), also known as a log-periodic array or log-periodic aerial, is a multi-element, directional antenna designed to operate over a wide band of frequencies. It was invented by Dwight Isbell and Raymond DuHamel at the University of Illinois in 1958.

The most common form of log-periodic antenna is the log-periodic dipole array or LPDA, The LPDA consists of a number of half-wave dipole driven elements of gradually increasing length, each consisting of a pair of metal rods. The dipoles are mounted close together in a line, connected in parallel to the feedline with alternating phase. Electrically, it simulates a series of two or three-element Yagi antennas connected together, each set tuned to a different frequency.

Log-periodic antenna, 400–4000 MHz

LPDA antennas look somewhat similar to Yagi antennas, in that they both consist of dipole rod elements mounted in a line along a support boom, but they work in very different ways. Adding elements to a Yagi increases its directionality, or gain, while adding elements to a LPDA increases its frequency response, or bandwidth.

One large application for LPDAs is in rooftop terrestrial television antennas, since they must have large bandwidth to cover the wide television bands of roughly 54–88 and 174–216 MHz in the VHF and 470–890 MHz in the UHF while also having high gain for adequate fringe reception. One widely used design for television reception combined a Yagi for UHF reception in front of a larger LPDA for VHF.

The LPDA normally consists of a series of half wave dipole "elements" each consisting of a pair of metal rods, positioned along a support boom lying along the antenna axis. The elements are spaced at intervals following a logarithmic function of the frequency, known as *d* or *sigma*. The successive elements gradually decrease in length along the boom. The relationship between the lengths is a function known as *tau*. *Sigma* and *tau* are the key design elements of the LPDA design. The radiation pattern of the antenna is unidirectional, with the main lobe along the axis of the boom, off the end with the shortest elements. Each dipole element is resonant at a wavelength approximately equal to twice its length. The bandwidth of the antenna, the frequency range over which it has maximum gain, is approximately between the resonant frequencies of the longest and shortest element.

Every element in the LPDA antenna is a driven element, that is, connected electrically to the feedline. A parallel wire transmission line usually runs along the central boom, and each successive element is connected in *opposite* phase to it. The feedline can often be seen zig-zagging across the support boom holding the elements. Another common construction method is to use two parallel central support booms that also acts as the transmission line, mounting the dipoles on the alternate booms. Other forms of the log-periodic design replace the dipoles with the transmission line itself, forming the log-periodic zig-zag antenna. Many other forms using the transmission wire as the active element also exist.

The Yagi and the LPDA designs look very similar at first glance, as they both consist of a number of dipole elements mounted along a support boom. The Yagi, however, has only a single driven element connected to the transmission line, usually the second one from the back of the array, the remaining elements are parasitic. The Yagi antenna differs from the LPDA in having a very narrow bandwidth.

In general terms, at any given frequency the log-periodic design operates somewhat similar to a three-element Yagi antenna; the dipole element closest to resonant at the operating frequency acts as a driven element, with the two adjacent elements on either side as director and reflector to increase the gain, the shorter element in front acting as a director and the longer element behind as a reflector. However, the system is somewhat more complex than that, and all the elements contribute to some degree, so the gain for any given frequency is higher than a Yagi of the same dimensions as any one section of the log-periodic. However, it should also be noted that a Yagi with the same number of elements as a log-periodic would have *far* higher gain, as all of those elements are improving the gain of a single driven element. In its use as a television antenna, it was common to combine a log-periodic design for VHF with a Yagi for UHF, with both halves being roughly equal in size. This resulted in much higher gain for UHF, typically on the order of 10 to 14 dB on the Yagi side and 6.5 dB for the log-periodic. But this extra gain was needed anyway in order to make up for a number of problems with UHF signals.

Log-periodic antenna, 250–2400 MHz

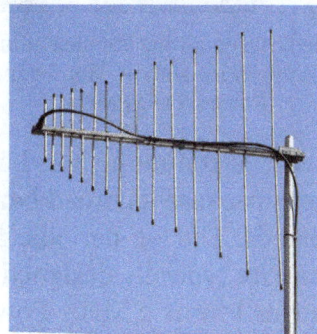

Log periodic mounted for vertical polarization, covers 140–470 MHz

It should be strictly noted that the log-periodic shape, according to the IEEE definition, does not align with broadband property for antennas. The broadband property of log-periodic antennas comes from its self-similarity. A planar log-periodic antenna can also be made self-complementary, such as logarithmic spiral antennas (which are not classified as log-periodic *per se* but among the frequency independent antennas that are also self-similar) or the log-periodic toothed design. Y. Mushiake found, for what he termed "the simplest self-complementary planar antenna," a driving point impedance of $\eta_o/2=188.4\ \Omega$ at frequencies well within its bandwidth limits.

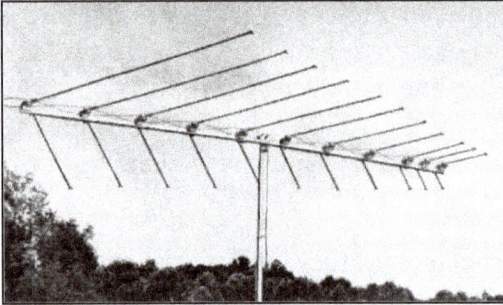

LP television antenna 1963. Covers 54–88 MHz and 174–218 MHz. Slanted elements were used because on the upper band they operate at the 3rd harmonic.

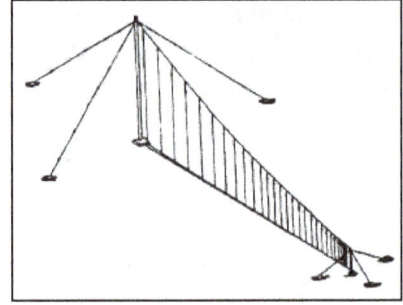

Wire Log-periodic monopole antenna.

Short Wave Broadcast Antennas

Wire log periodic transmitting antenna at international shortwave broadcasting station, Moosbrunn, Austria. Covers 6.1–23 MHz

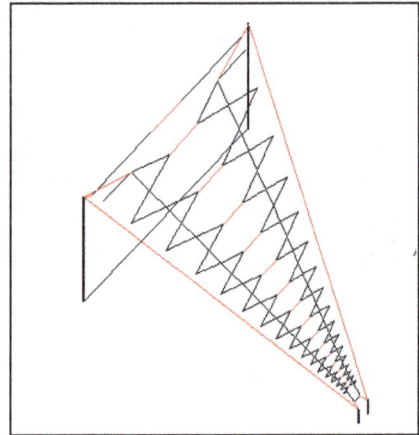

Diagram of a zig-zag shortwave LPA antenna, black shows metallic conductors, red shows insulating supports

The log periodic is commonly used as a transmitting antenna in high power shortwave broadcasting stations because its broad bandwidth allows a single antenna to transmit on frequencies in multiple bands. The log-periodic zig-zag design with up to 16 sections has been used. These large antennas are typically designed to cover 6 to 26 MHz but even larger ones have been built which operate as low as 2 MHz. Power ratings are available up to 500 kW. Instead of the elements being driven in parallel, attached to a central transmission line, the elements are driven in series, adjacent elements connected at the outer edges. The antenna shown here would have about 14 dBi gain. An antenna array consisting of two such antennas, one above the other and driven in phase has a gain of up to 17 dBi. Being log-periodic, the antenna's main characteristics (radiation pattern, gain, driving point impedance) are almost constant over its entire frequency range, with the match to a 300 Ω feed line achieving a standing wave ratio of better than 2:1 over that range.

BROAD-SIDE ARRAY

The antenna array in its simplest form, having a number of elements of equal size, equally spaced along a straight line or axis, forming collinear points, with all dipoles in the same phase, from the same source together form the broad side array.

Frequency Range

The frequency range, in which the collinear array antennas operate is around 30 MHz to 3GHz which belong to the VHF and UHF bands.

Construction and Working of Broad-side Array

According to the standard definition, "An arrangement in which the principal direction of radiation is perpendicular to the array axis and also to the plane containing the array element" is termed as the broad side array. Hence, the radiation pattern of the antenna is perpendicular to the axis on which the array exists.

The following diagram shows the broad side array, in front view and side view, respectively.

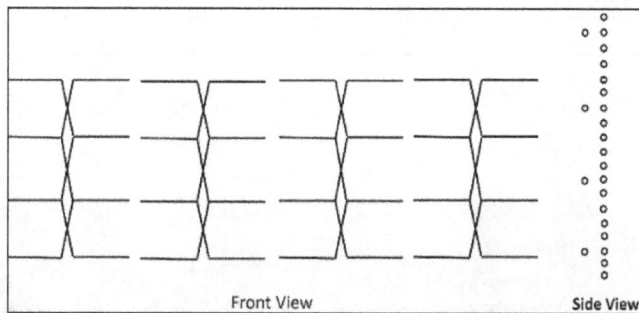

Front View   Side View

The broad side array is strongly directional at right angles to the plane of the array. However, the radiation in the plane will be very less because of the cancellation in the direction joining the center.

The figure of broad side array with $\lambda/4$ spacing is shown below.

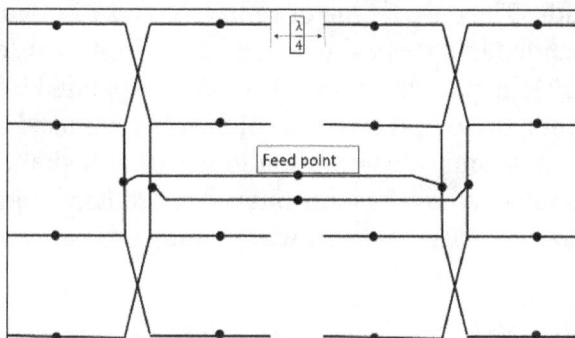

Typical antenna lengths in the broad side array are from 2 to 10 wavelengths. Typical spacings are $\lambda/2$ or λ. The feed points of the dipoles are joined as shown in the figure.

Radiation Pattern

The radiation pattern of this antenna is bi-directional and right angles to the plane. The beam is very narrow with high gain.

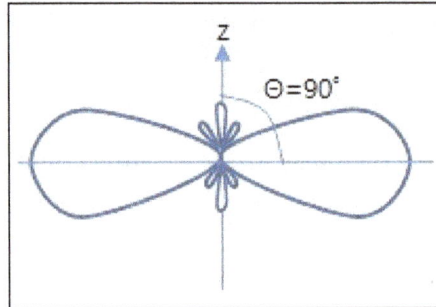

The above figure shows the radiation pattern of the broad side array. The beam is a bit wider and minor lobes are much reduced in this.

CURTAIN ARRAY

Curtain arrays at Radio Free Europe transmitter site, Biblis, Germany

Curtain arrays are a class of large multielement directional wire radio transmitting antennas, used in the shortwave radio bands. They are a type of reflective array antenna, consisting of multiple wire dipole antennas, suspended in a vertical plane, often in front of a "curtain" reflector made of a flat vertical screen of many long parallel wires. These are suspended by support wires strung between pairs of tall steel towers, up to 300 ft (90 m) high. They are used for long-distance skywave (or *skip*) transmission; they transmit a beam of radio waves at a shallow angle into the sky just above the horizon, which is reflected by the ionosphere back to Earth beyond the horizon. Curtain antennas are mostly used by international short wave radio stations to broadcast to large areas at transcontinental distances.

Because of their powerful directional characteristics, curtain arrays are often used in government propaganda radio stations to beam propaganda broadcasts over national borders into other nations. For example, curtain arrays were used by Radio Free Europe and Radio Liberty to broadcast into Eastern Europe.

Curtain array at international shortwave broadcasting station, Moosbrunn, Austria. It consists of 4 columns of horizontal wire dipoles, suspended in front of a wire screen. The vertical parallel wire feedlines to each column of dipoles are visible. The entire antenna is mounted on a rotating truss structure, allowing it to be pointed in different directions.

The driven elements are usually half-wave dipoles, fed in phase, mounted in a plane $\frac{1}{4}$ wavelength in front of the reflector plane. The reflector wires are oriented parallel to the dipoles. The dipoles may be vertical, radiating in vertical polarization, but are most often horizontal, because horizontally polarized waves are less absorbed by earth reflections. The lowest row of dipoles are mounted more than $\frac{1}{2}$ wavelength above the ground, to prevent ground reflections from interfering with the radiation pattern. This allows most of the radiation to be concentrated in a narrow main lobe aimed a few degrees above the horizon, which is ideal for skywave transmission. A curtain array may have a gain of 20 dB greater than a simple dipole antenna. Because of the strict phase requirements, earlier curtain arrays had a narrow bandwidth, but modern curtain arrays can be built with a bandwidth of up to 2:1, allowing them to cover several shortwave bands.

Rather than feeding each dipole at its center, which requires a "tree" transmission line structure with complicated impedance matching, multiple dipoles are often connected in series to make an elaborate folded dipole structure which can be fed at a single point.

In order to allow the beam to be steered, sometimes the entire array is suspended by cantilever arms from a single large tower which can be rotated. Alternatively, some modern versions are constructed as phased arrays in which the beam can be slewed electronically, without moving the antenna. Each dipole or group of dipoles is fed through an electronically adjustable phase shifter, implemented either by passive networks of capacitors and inductors which can be switched in and out, or by separate output RF amplifiers. Adding a constant phase shift between adjacent horizontal dipoles allows the direction of the beam to be slewed in azimuth up to +/- 30° without losing its radiation pattern.

Three-array Systems

Transmission system are optimized for geopolitical reasons. Geopolitical necessity leads some international broadcasters to occasionally use three separate antenna arrays: highband and midband, as well as lowband HRS curtains.

Using three curtain arrays to cover the HF broadcasting spectrum creates a highly optimized HF transmission system, but three or more curtain arrays can be costly to build and maintain, and no new HF relay stations have been built since the mid-1990s. The modern HRS antenna design has a long lifespan, however, so existing HRS shortwave transmission systems built before 1992 will likely remain available for some time.

Nomenclature

Since 1984 the CCIR has created a standardised nomenclature for describing curtain antennas, consisting of 1 to 4 letters followed by three numbers:

First letter: Indicates the orientation of the dipoles in the array.

- "H" indicates the dipoles are oriented horizontally, so the antenna radiates horizontally polarized radio waves.

- "V" indicates the dipoles are oriented vertically, so the antenna radiates vertically polarized radio waves.

Former Radio France Internationale (RFI) Issoudun Relay station feeders and curtain arrays.

Second letter (if present): Indicates whether the antenna has a reflector.

- "R" indicates that there is a simple (passive) reflector on one side of the array, so the antenna radiates a single beam.

- "RR" indicates that the array has some kind of "reversible reflector", so the direction of the beam can be switched 180°. Very few of this type have ever been built. RCI Sackville in Canada may have 2 HRRS type antennas—perhaps the only ones in North America.

- If "R" and "RR" are missing, the antenna has no reflector, so the dipole array will radiate its energy in two beams in both directions perpendicular to its plane, 180° apart.

Third letter (if present):

- "S" indicates that the array is steerable.

Following the letters come three numbers:

"x/y/z".

"x" and "y" specifies the dimensions of the rectangular array of dipoles, while "z" gives the height above the ground of the bottom of the array:

- "x" (an integer) is the number of dipoles in horizontal rows.

- "y" (an integer) is the number of vertical columns of rows (dipoles).

- "z" (a decimal fraction) is the height above ground in wavelengths of the lowest row of dipoles in the array.

For example, a "HRS 4/5/0.5" curtain antenna has a rectangular array of 20 dipoles, 4 dipoles wide and 5 dipoles high, with the lowest row being half a wavelength off the ground, and a flat reflector behind it, and the direction of the beam can be slewed. An HRS 4/4/0.5 slewable antenna with 16 dipoles is one of the standard types of array seen at shortwave broadcast stations worldwide.

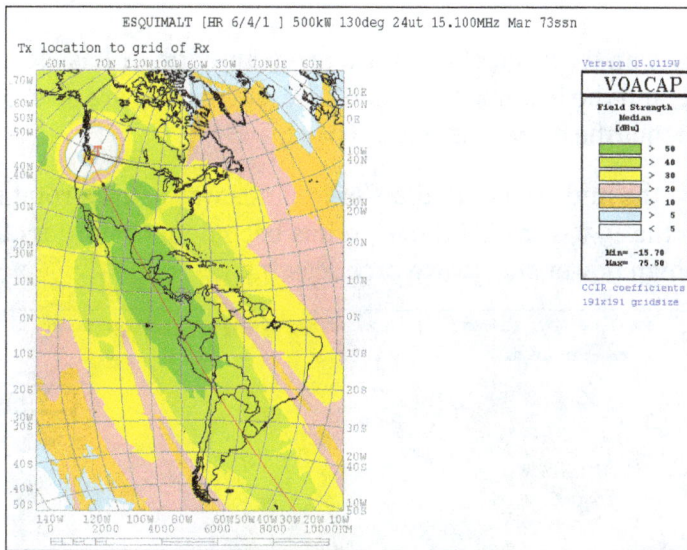

Simulated radiation pattern of a 15.1 MHz HR 6/4/1 curtain antenna (24 horizontal dipoles organized in 4 rows of 6 elements each, in front of a reflector), driven by a 500 kW transmitter. The transmitter is located in Seattle and the pattern covers Central America and parts of South America, showing the long distances achieved with this antenna. The main lobe of the pattern is flanked by two sidelobes, which appear curved due to the global map projection.

HRS

An HRS type antenna is basically a rectangular array of conventional dipole antennas strung between supporting towers. In the simplest case, each dipole separated from the next by 1/2 λ vertically, and the centres of each dipole are spaced 1 λ apart horizontally. Again, in the simplest case (for a broadside beam), all dipoles are driven in phase with each other and with equal power. Radiation is concentrated broadside to the curtain.

Behind the array of dipoles, typically about 1/3 λ away there will be a "reflector" consisting of many parallel wires in the same orientation as the dipoles. If this was not present, the curtain would radiate equally forward and backward.

HRS nomenclature:

- HRS antennas of type HRS 1/1/z are undefined as such (such a thing would consist of just a single dipole).

- HRS antennas of type HRS 1/2/z and 2/1/z exist, but see little practical use in shortwave broadcasting. VHF and UHF repeaters for FM radio or television in the UK quite often employ a pair of horizontal dipoles (or short yagis) one above the other (i.e. HRS 1/2/z) to concentrate transmission power in the vertical plane.

- The Russian Duga Over The Horizon Radar may have used an antenna of type HRS 32/16/0.75 (estimated – not verified), with potential directional ERP in the gigawatt range.

HRS Antenna

The HRS type antenna is an example of a curtain array antenna. It has Horizontal dipoles with a Reflector behind them, and the beam is Steerable. These antennas are also known as "HRRS" (for a Reversible Reflector), but the extra R is seldom used.

However, as far back as the mid-1930s, Radio Netherlands was using a rotatable HRS antenna for global coverage. Since the 1950s the HRS design has become more or less the standard for long distance (> 1000 km) high power shortwave broadcasting.

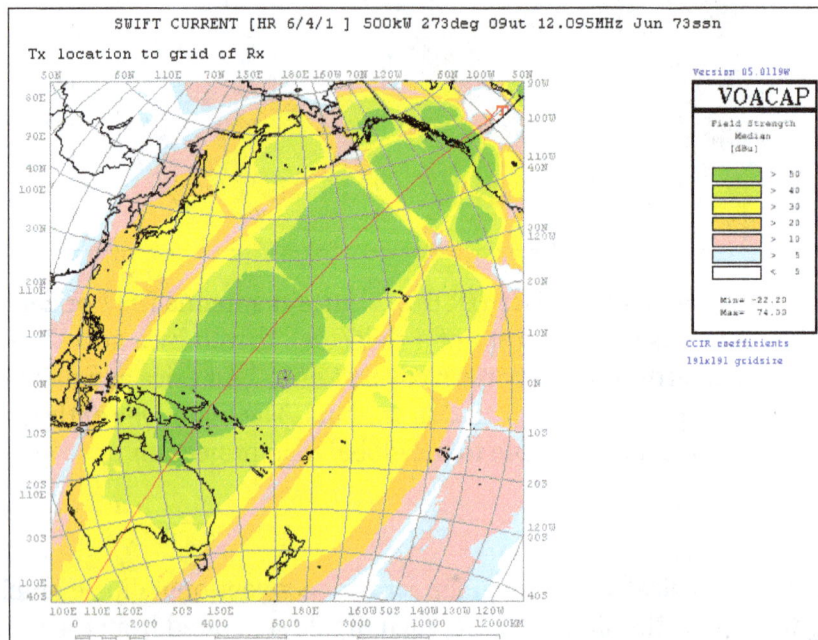

Example of a simulated HRS antenna radiation pattern from a shortwave relay station in Canada.
It consists of a main lobe with two major sidelobes. The sidelobes look curved because of the map projection.

Steering

If there is an "S" in the antenna's designation, it is a steerable design. Following the ITU-recommandation, it might be called 'slewable design'. This might be achieved electronically by adjustment of the electrical wave phases of the signals fed to the columns of dipole antenna elements, or physically by mounting the antenna array on a large rotating mechanism. An example of this can be seen at NRK Kvitsøy, where a circular railway carries a pair of wheeled platforms, each of which supports a tower at opposite ends of a diameter-arm. The curtain antenna array is suspended between the towers and rotates with them as the towers go around the circular railway. Another physical rotation technique is employed by the ALLISS system where the entire array is built around a central rotatable tower of great strength.

ALLISS antenna as viewed underneath

Electrically slewed antenna arrays can usually be aimed in the range of ±30° from the antenna's physical direction while mechanically rotated arrays can accommodate a full 360°. Electrical slewing is typically done in the horizontal plane, with some adjustment being possible in the vertical plane.

Azimuth Beamwidth

- For a 2-wide dipole array, the beamwidth is around 50°
- For a 3-wide dipole array, the beamwidth is around 40°
- For a 4-wide dipole array, the beamwidth is around 30°

Vertical Launch Angle

The number of dipole rows and the height of the lowest element above ground determine the elevation angle and consequently the distance to the service area.

- A 2-row high array has a typical takeoff angle of 20°, is most commonly used for medium range communications.
- A 4-row high array has a typical takeoff angle of 10°, is most commonly used for long range communications.

- A 6-row array is similar to a 4-row, but can achieve 5° to 10° takeoff angles, can be used in shortwave communications circuits of 12000 km, and is highly directional.

Note that it is possible for details of the antenna site to wreak havoc with the designers plans such that takeoff angle and matching may be adversely affected.

END-FIRE ARRAY

The physical arrangement of end-fire array is same as that of the broad side array. The magnitude of currents in each element is same, but there is a phase difference between these currents. This induction of energy differs in each element, which can be understood by the following diagram.

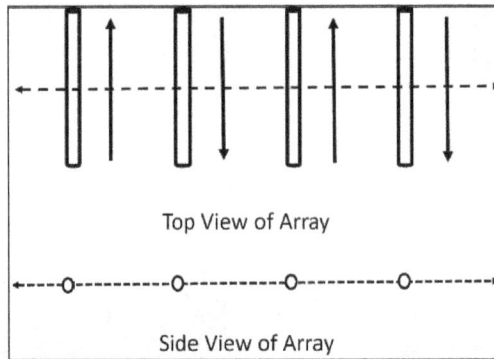

Top View of Array

Side View of Array

The above figure shows the end-fire array in top and side views respectively.

There is no radiation in the right angles to the plane of the array because of cancellation. The first and third elements are fed out of phase and therefore cancel each other's radiation. Similarly, second and fourth are fed out of phase, to get cancelled.

The usual dipole spacing will be $\lambda/4$ or $3\lambda/4$. This arrangement not only helps to avoid the radiation perpendicular to the antenna plane, but also helps the radiated energy get diverted to the direction of radiation of the whole array. Hence, the minor lobes are avoided and the directivity is increased. The beam becomes narrower with the increased elements.

Radiation Pattern

The Radiation pattern of end-fire array is uni-directional. A major lobe occurs at one end, where maximum radiation is present, while the minor lobes represent the losses.

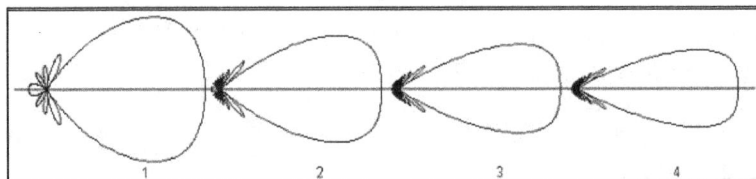

The figure explains the radiation pattern of an end-fire array. Figure is the radiation pattern for a single array, while figures above represent the radiation pattern for multiple arrays.

End-fire Array vs. Broad Side Array

We have studied both the arrays. Let us try to compare the end-fire and broad side arrays, along with their characteristics.

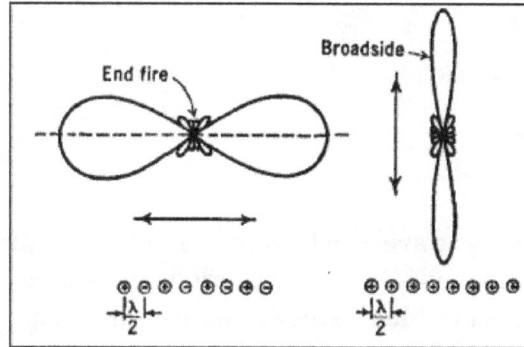

The figure illustrates the radiation pattern of end-fire array and broad side array:

- Both, the end fire array and broad side array, are linear and are resonant, as they consist of resonant elements.

- Due to resonance, both the arrays display narrower beam and high directivity.

- Both of these arrays are used in transmission purposes.

- Neither of them is used for reception, because the necessity of covering a range of frequencies is needed for any kind of reception.

References

- Main, arrays: antenna-theory.com, Retrieved 29 May 2019

- Raut, s. (july 2014). "wideband printed bowtie array for spectrum monitoring". 2014 ieee antennas and propagation society international symposium (apsursi). Antennas and propagation society international symposium. Pp. 235–236. Doi:10.1109/aps.2014.6904449. Isbn 978-1-4799-3540-6

- Antenna-theory-parasitic-arrays, antenna-theory: tutorialspoint.com, Retrieved 21, August 2019

- Griffith, b. Whitfield (2000). Radio-electronic transmission fundamentals, 2nd ed. Scitech publishing. P. 477. Isbn 1884932134

- "Aegis weapon system mk-7". Jane's information group. 2001-04-25. Archived from the original on 1 july 2006. Retrieved 10 august2006

- Josefsson, patrik persson, lars josefsson, patrik persson; patrik persson (2006). Conformal array antenna theory and design. Usa: john wiley and sons. Pp. 1–9. Isbn 0-471-46584-4

- Antenna-theory-end-fire-array, antenna-theory: tutorialspoint.com, Retrieved 13 February, 2019

6

Wave Propagation

The different ways in which waves travel can be referred to as wave propagation. Sky wave propagation, space wave propagation, ionospheric propagation, ground wave propagation, radio propagation, etc. are some of the ways in which waves propagate. This chapter has been carefully written to provide an easy understanding of wave propagation.

ELECTROMAGNETIC WAVES

Electromagnetic radiation is a form of energy that is produced by oscillating electric and magnetic disturbance, or by the movement of electrically charged particles traveling through a vacuum or matter. The electric and magnetic fields come at right angles to each other and combined wave moves perpendicular to both magnetic and electric oscillating fields thus the disturbance. Electron radiation is released as photons, which are bundles of light energy that travel at the speed of light as quantized harmonic waves. This energy is then grouped into categories based on its wavelength into the electromagnetic spectrum. These electric and magnetic waves travel perpendicular to each other and have certain characteristics, including amplitude, wavelength, and frequency.

General Properties of all electromagnetic radiation:

1. Electromagnetic radiation can travel through empty space. Most other types of waves must travel through some sort of substance. For example, sound waves need either a gas, solid, or liquid to pass through in order to be heard.

2. The speed of light is always a constant. (Speed of light : 2.99792458×10^8 m s^{-1})

3. Wavelengths are measured between the distances of either crests or troughs. It is usually characterized by the Greek symbol λ.

Waves and their Characteristics

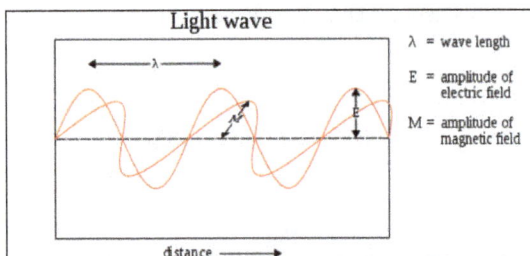

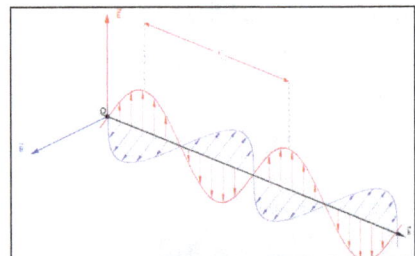

Electromagnetic Waves.

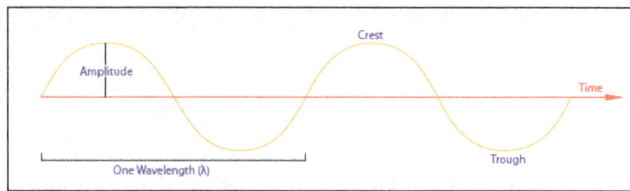

An EM Wave.

Amplitude

Amplitude is the distance from the maximum vertical displacement of the wave to the middle of the wave. This measures the magnitude of oscillation of a particular wave. In short, the amplitude is basically the height of the wave. Larger amplitude means higher energy and lower amplitude means lower energy. Amplitude is important because it tells you the intensity or brightness of a wave in comparison with other waves.

Wavelength

Wavelength (λ) is the distance of one full cycle of the oscillation. Longer wavelength waves such as radio waves carry low energy; this is why we can listen to the radio without any harmful consequences. Shorter wavelength waves such as x-rays carry higher energy that can be hazardous to our health. Consequently lead aprons are worn to protect our bodies from harmful radiation when we undergo x-rays. This wavelength frequently relationship is characterized by:

$$c = \lambda v$$

where:

- c is the speed of light,

- λ is wavelength, and

- v is frequency.

Shorter wavelength means greater frequency, and greater frequency means higher energy. Wavelengths are important in that they tell one what type of wave one is dealing with.

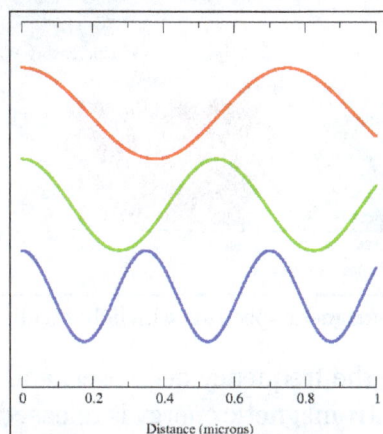

Different Wavelengths and Frequencies.

Remember, Wavelength tells you the type of light and Amplitude tells you about the intensity of the light.

Frequency

Frequency is defined as the number of cycles per second, and is expressed as sec⁻¹ or Hertz (Hz). Frequency is directly proportional to energy and can be express as:

$$E = h\nu$$

where:

- E is energy,

- h is Planck's constant, (h= 6.62607 x 10⁻³⁴ J), and

- ν is frequency.

Period

Period (T) is the amount of time a wave takes to travel one wavelength; it is measured in seconds (s).

Velocity

The velocity of wave in general is expressed as:

$$velocity = \lambda \nu$$

For Electromagnetic wave, the velocity in vacuum is $2.99 \times 10^8 \ m/s$ or $186,282$ miles/second.

Electromagnetic Spectrum

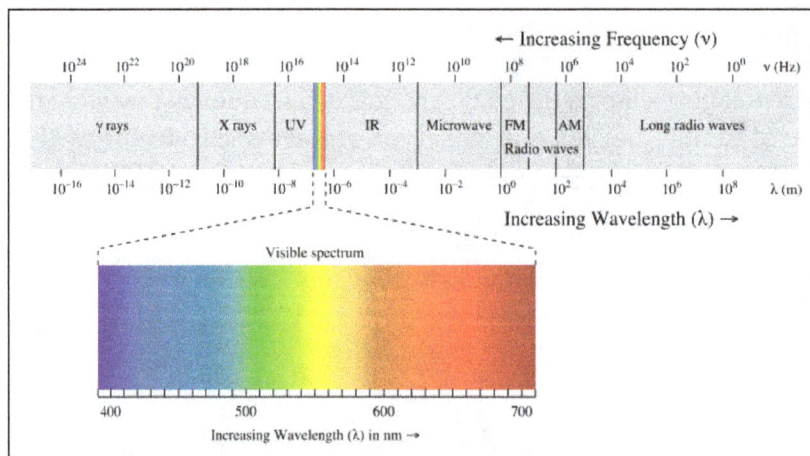

Electromagnetic spectrum with light highlighted.

As a wave's wavelength increases, the frequency decreases, and as wave's wavelength decreases, the frequency increases. When electromagnetic energy is released as the energy level increases, the wavelength decreases and frequency decreases. Thus, electromagnetic radiation is then grouped into categories based on its wavelength or frequency into the electromagnetic spectrum. The different types of electromagnetic radiation shown in the electromagnetic spectrum consists of radio waves, microwaves, infrared waves, visible light, ultraviolet radiation, X-rays, and gamma rays. The part of the electromagnetic spectrum that we are able to see is the visible light spectrum.

Electromagnetic Spectrum with Radiation Types.

Radiation Types

Radio Waves are approximately 10^3 m in wavelength. As the name implies, radio waves are transmitted by radio broadcasts, TV broadcasts, and even cell phones. Radio waves have the lowest energy levels. Radio waves are used in remote sensing, where hydrogen gas in space releases radio energy with a low frequency and is collected as radio waves. They are also used in radar systems, where they release radio energy and collect the bounced energy back. Especially useful in weather, radar systems are used to can illustrate maps of the surface of the Earth and predict weather patterns since radio energy easily breaks through the atmosphere.

Microwaves can be used to broadcast information through space, as well as warm food. They are also used in remote sensing in which microwaves are released and bounced back to collect information on their reflections.

This picture represents a snap shot in mid-infrared light.

Microwaves can be measured in centimeters. They are good for transmitting information because the energy can go through substances such as clouds and light rain. Short microwaves are sometimes used in Doppler radars to predict weather forecasts.

Infrared radiation can be released as heat or thermal energy. It can also be bounced back, which is called near infrared because of its similarities with visible light energy. Infrared Radiation is most commonly used in remote sensing as infrared sensors collect thermal energy, providing us with weather conditions.

Visible Light is the only part of the electromagnetic spectrum that humans can see with an unaided eye. This part of the spectrum includes a range of different colors that all represent a particular wavelength. Rainbows are formed in this way; light passes through matter in which it is absorbed or reflected based on its wavelength. Thus, some colors are reflected more than other, leading to the creation of a rainbow.

Color Region	Wavelength (nm)
Violet	380-435
Blue	435-500
Cyan	500-520
Green	520-565
Yellow	565-590
Orange	590-625
Red	625-740

The color regions of the Visible Spectrum.

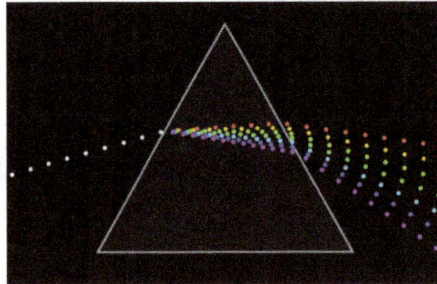

Dispersion of Light Through A Prism.

Ultraviolet, Radiation, X-Rays, and Gamma Rays are all related to events occurring in space. UV radiation is most commonly known because of its severe effects on the skin from the sun, leading to cancer. X-rays are used to produce medical images of the body. Gamma Rays can used in chemotherapy in order to rid of tumors in a body since it has such a high energy level. The shortest waves, Gamma rays, are approximately 10^{-12} m in wavelength. Out this huge spectrum, the human eyes can only detect waves from 390 nm to 780 nm.

Equations of Waves

The mathematical description of a wave is:

$$y = A\sin(kx - \omega t)$$

where A is the amplitude, k is the wave number, x is the displacement on the x-axis.

$$k = \frac{2\pi}{\lambda}$$

where λ is the wavelength. Angular frequency described as:

$$\omega = 2\pi v = \frac{2\pi}{T}$$

where v is frequency and period (T) is the amount of time for the wave to travel one wavelength.

Interference

An important property of waves is the ability to combine with other waves. There are two type of interference: constructive and destructive. Constructive interference occurs when two or more waves are in phase and and their displacements add to produce a higher amplitude. On the contrary, destructive interference occurs when two or more waves are out of phase and their displacements negate each other to produce lower amplitude.

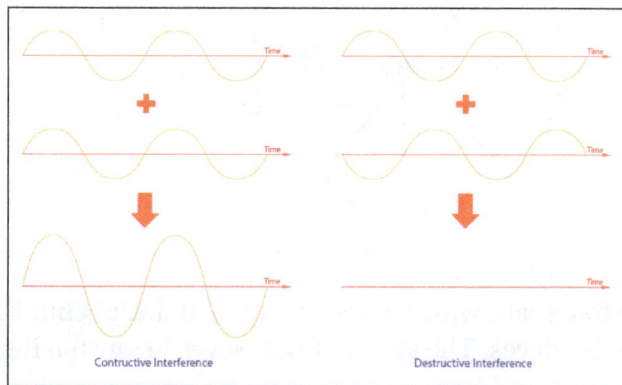

Constructive and Destructive Interference

Interference can be demonstrated effectively through the double slit experiment. This experiment consists of a light source pointing toward a plate with one slit and a second plate with two slits. As the light travels through the slits, we notice bands of alternating intensity on the wall behind the second plate. The banding in the middle is the most intense because the two waves are perfectly in phase at that point and thus constructively interfere. The dark bands are caused by out of phase waves which result in destructive interference. This is why you observe nodes on figure. In a similar way, if electrons are used instead of light, electrons will be represented both as waves and particles.

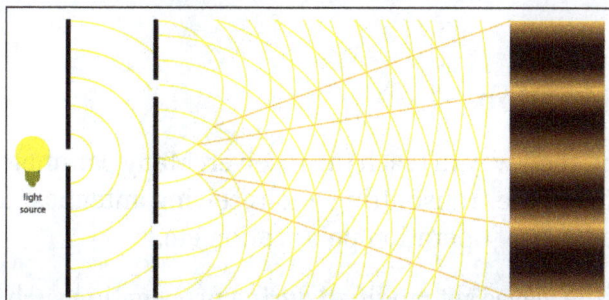

WAVE PROPAGATION

Electromagnetic Waves are generated by the radiated power from the current carrying conductor. In conductors, a part of the generated power escapes and propagates into free space in the form of Electromagnetic wave, which has a time-varying electrical field, magnetic field, and direction of propagation orthogonal to each other.

Radiated from an isotropic transmitter, these wave travels through different paths to reach the receiver. The path taken by the wave to travel from the transmitter and reach the receiver is known as Wave Propagation.

When the isotropic radiator is used for transmission of EM waves we get spherical wavefronts as shown in the figure because it radiates EM waves uniformly and equally in all directions. Here the center of the sphere is the radiator while the radius of the sphere is R. Clearly, all the points at the distance R, lying on the surface of the sphere have equal power densities.

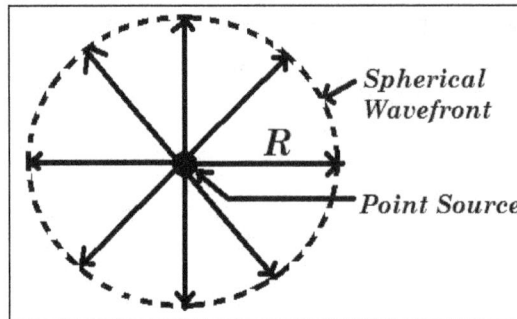

Spherical Wavefront.

The E waves travel in the free space with the velocity of light .i.e. c = But EM waves travel through another medium the speed reduces. The speed of EM waves in any medium other than free space is given by,

$$v = \frac{c}{\sqrt{\epsilon_r}}$$

where c is the velocity of light and is relative permittivity of the medium.

EM waves transmit energy by absorption and re-emission of wave energy by the atoms in the medium. The atoms absorb the wave energy, undergo vibrations and pass the energy by re-emission of EM of the same frequency. The optical density of the medium affects the propagation of EM waves.

Wave Propagation Equation

Waves take many routes on their way to reach the receiver. Many parameters decide the path taken by the wave such as heights of the transmitting and receiving antennas, the angle of launching at transmitting end, the frequency of operation polarization etc.

Many of the properties of the waves get modified during propagation such as reflection, refraction, diffraction etc. due to the variation of parameters of propagating media like conductivity, permittivity, permeability, and characteristics of obstructing objects.

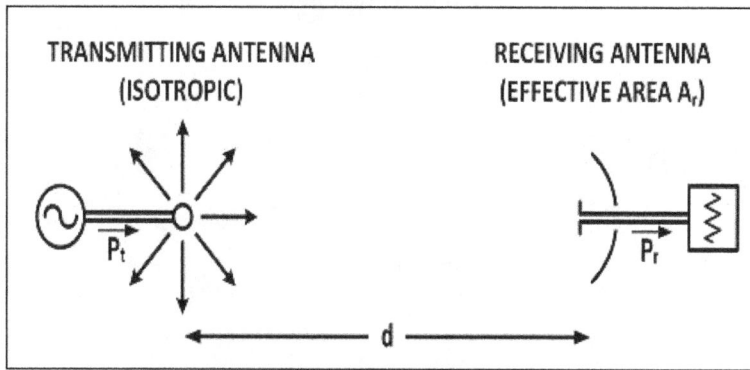

TRANSMITTING ANTENNA
(ISOTROPIC)

RECEIVING ANTENNA
(EFFECTIVE AREA A$_r$)

Friis Free Space Radio Circuit

Generally, when power is radiated in the free space, the wave energy may be radiated or absorbed by the objects in the medium. So while transmitting a wave through a medium it is essential to calculate the loss that can be caused to the wave. This loss is called Radio transmission loss, which is based on inverse square law of optics and is calculated as the ratio of the radiated power to the received power.

As we know that when an isotropic transmitter is used, power is distributed equally, average power can be expressed in terms of radiated power as,

$$P_{avg} = \frac{\dfrac{P_{rad}}{4\pi r^2} \, W}{m^2}$$

The directivity of a test antenna is given by,

$$\left(G_{D\,max}\right)_t = \frac{P_{d\,max}}{\dfrac{P_{rad}}{4\pi r^2}}$$

$$\therefore P_{d\,max} = G_{D\,max} \cdot \frac{P_{rad}}{4\pi r^2}$$

Assume that the receiving antenna receives all the generated power from the radio waves without any loss. Let be the maximum power received by the receiver antenna under a matched load condition. When is the effective aperture of the receiving antenna, we can write as,

$$P_{rec} = P_{d\,max} \left(A_e\right)_r$$

$$\therefore P_{rec} = \left(G_{D\,max}\right)_t \cdot \frac{P_{rad}}{4\pi r^2} \left(A_e\right)_r$$

In general, the directivity and effective aperture area for any antenna is related as,

$$G_{D\,max} = \frac{4\pi}{\lambda^2} \left(A_e\right)$$

Let be the directivity of the receiving antenna. Then,

Let $\left(G_{D\max}\right)_r$ be the directivity of the receiving antenna. Then,

$$\left(G_{D\max}\right)_r = \frac{4\pi}{\lambda^2}(A_e)_r$$

$$\therefore (A_e)_r = \frac{\lambda^2}{4\pi}\left(G_{D\max}\right)_r$$

Substituting the value in $P_{d\max} = G_{D\max} \cdot \dfrac{P_{rad}}{4\pi r^2}$ we get,

$$P_{rec} = \left(G_{D\max}\right)_t \cdot \frac{P_{rad}}{4\pi r^2}[\frac{\lambda^2}{4\pi}\left(G_{D\max}\right)_r]$$

$$\therefore \frac{P_{rec}}{P_{rad}} = \left(G_{D\max}\right)_t \left(G_{D\max}\right)_r \left(\frac{\lambda}{4\pi r}\right)^2$$

This equation is known as the Fundamental Equation for free space Propagation, also known as Friss free space equation. The factor $(\lambda/4\pi r)^2$ is called free space path loss which indicates the loss of the signal. Path loss can be expressed as,

$$P_{Loss} = 10\log_{10}(\frac{4\pi r}{\lambda})^2\, dB$$

We can express the equation $\dfrac{P_{rec}}{P_{rad}} = \left(G_{D\max}\right)_t \left(G_{D\max}\right)_r \left(\dfrac{\lambda}{4\pi r}\right)^2$ in dB as,

$$10\log_{10}\left(\frac{P_{rec}}{P_{rad}}\right) = 10\log_{10}{(G_{D\max})_t} + 10\log_{10}{(G_{D\max})_r} + 10\log_{10}\left[\left(\frac{\lambda}{4\pi r}\right)^2\right]$$

Received power can be expressed as,

$$P_{rec(dB)} = P_{rad(dB)} + \left(G_{D\max}\right)_{r(dB)} + \left(G_{D\max}\right)_{t(dB)} - L_{s(dB)}$$

Where $L_{s(dB)} = 10\log_{10}\left[\left(\dfrac{\lambda}{4\pi r}\right)^2\right] = 20\log_{10}(\dfrac{\lambda}{4\pi r})$

Which, on simplification is given as,

$$L_{s(dB)} = 32.45 + 20\log_{10} r + 20\log_{10} f,$$

Here distance r is expressed in kilometer while frequency f is expressed in MHz. This indicates loss due to wave spreading taking place when it propagates out of the source.

RADIO WAVES

Radio waves are a type of electromagnetic radiation with wavelengths in the electromagnetic spectrum longer than infrared light. Radio waves have frequencies as high as 300 gigahertz (GHz) to as low as 30 hertz (Hz). At 300 GHz, the corresponding wavelength is 1 mm, and at 30 Hz is 10,000 km. Like all other electromagnetic waves, radio waves travel at the speed of light in vacuum. They are generated by electric charges undergoing acceleration, such as time varying electric currents. Naturally occurring radio waves are emitted by lightning and astronomical objects.

Radio waves are generated artificially by transmitters and received by radio receivers, using antennas. Radio waves are very widely used in modern technology for fixed and mobile radio communication, broadcasting, radar and other navigation systems, communications satellites, wireless computer networks and many other applications. Different frequencies of radio waves have different propagation characteristics in the Earth's atmosphere; long waves can diffract around obstacles like mountains and follow the contour of the earth (ground waves), shorter waves can reflect off the ionosphere and return to earth beyond the horizon (skywaves), while much shorter wavelengths bend or diffract very little and travel on a line of sight, so their propagation distances are limited to the visual horizon.

To prevent interference between different users, the artificial generation and use of radio waves is strictly regulated by law, coordinated by an international body called the International Telecommunications Union (ITU), which defines radio waves as "electromagnetic waves of frequencies arbitrarily lower than 3 000 GHz, propagated in space without artificial guide". The radio spectrum is divided into a number of radio bands on the basis of frequency, allocated to different uses.

Speed, Wavelength and Frequency

Radio waves in a vacuum travel at the speed of light. When passing through a material medium, they are slowed according to that object's permeability and permittivity. Air is thin enough that in the Earth's atmosphere radio waves travel very close to the speed of light.

The wavelength is the distance from one peak of the wave's electric field (wave's peak/crest) to the next, and is inversely proportional to the frequency of the wave. The distance a radio wave travels in one second, in a vacuum, is 299,792,458 meters (983,571,056 ft) which is the wavelength of a 1 hertz radio signal. A 1 megahertz radio signal has a wavelength of 299.8 meters (984 ft).

Propagation

The study of radio propagation, how radio waves move in free space and over the surface of the Earth, is vitally important in the design of practical radio systems. Radio waves passing through different environments experience reflection, refraction, polarization, diffraction, and absorption. Different frequencies experience different combinations of these phenomena in the Earth's atmosphere, making certain radio bands more useful for specific purposes than others. Practical radio systems mainly use three different techniques of radio propagation to communicate:

- Line of sight: This refers to radio waves that travel in a straight line from the transmitting antenna to the receiving antenna. It does not necessarily require a cleared sight path; at lower frequencies radio waves can pass through buildings, foliage and other obstructions. This is the only method of propagation possible at frequencies above 30 MHz. On the surface of the Earth, line of sight propagation is limited by the visual horizon to about 64 km (40 mi). This is the method used by cell phones, FM, television broadcasting and radar. By using dish antennas to transmit beams of microwaves, point-to-point microwave relay links transmit telephone and television signals over long distances up to the visual horizon. Ground stations can communicate with satellites and spacecraft billions of miles from Earth.

  ○ Indirect propagation: Radio waves can reach points beyond the line-of-sight by *diffraction* and *reflection*. Diffraction allows a radio wave to bend around obstructions such as a building edge, a vehicle, or a turn in a hall. Radio waves also partially reflect from surfaces such as walls, floors, ceilings, vehicles and the ground. These propagation methods occur in short range radio communication systems such as cell phones, cordless phones, walkie-talkies, and wireless networks. A drawback of this mode is *multipath propagation*, in which radio waves travel from the transmitting to the receiving antenna via multiple paths. The waves interfere, often causing fading and other reception problems.

- Ground waves: At lower frequencies below 2 MHz, in the medium wave and longwave bands, due to diffraction vertically polarized radio waves can bend over hills and mountains, and propagate beyond the horizon, traveling as surface waves which follow the contour of the Earth. This allows mediumwave and longwave broadcasting stations to have coverage areas beyond the horizon, out to hundreds of miles. As the frequency drops, the losses decrease and the achievable range increases. Military very low frequency (VLF) and extremely low frequency (ELF) communication systems can communicate over most of the Earth, and with submarines hundreds of feet underwater.

- Skywaves: At medium wave and shortwave wavelengths, radio waves reflect off conductive layers of charged particles (ions) in a part of the atmosphere called the ionosphere. So radio waves directed at an angle into the sky can return to Earth beyond the horizon; this is called "skip" or "skywave" propagation. By using multiple skips communication at intercontinental distances can be achieved. Skywave propagation is variable and dependent on atmospheric conditions; it is most reliable at night and in the winter. Widely used during the first half of the 20th century, due to its unreliability skywave communication has mostly been abandoned. Remaining uses are by military over-the-horizon (OTH) radar systems, by some automated systems, by radio amateurs, and by shortwave broadcasting stations to broadcast to other countries.

Radio Communication

In radio communication systems, information is carried across space using radio waves. At the sending end, the information to be sent, in the form of a time-varying electrical signal, is applied to a radio transmitter. The information signal can be an audio signal representing sound from a microphone, a video signal representing moving images from a video camera, or a digital signal representing data from a computer. In the transmitter, an electronic oscillator generates an alternating current oscillating at a radio frequency, called the *carrier wave* because it serves to "carry" the information through the air. The information signal is used to modulate the carrier, altering some aspect of it, "piggybacking" the information on the carrier. The modulated carrier is amplified and applied to an antenna. The oscillating current pushes the electrons in the antenna back and forth, creating oscillating electric and magnetic fields, which radiate the energy away from the antenna as radio waves. The radio waves carry the information to the receiver location.

At the receiver, the oscillating electric and magnetic fields of the incoming radio wave push the electrons in the receiving antenna back and forth, creating a tiny oscillating voltage which is a weaker replica of the current in the transmitting antenna. This voltage is applied to the radio receiver, which extracts the information signal. The receiver first uses a bandpass filter to separate the desired radio station's radio signal from all the other radio signals picked up by the antenna, then amplifies the signal so it is stronger, then finally extracts the information-bearing modulation signal in a demodulator. The recovered signal is sent to a loudspeaker or earphone to produce sound, or a television display screen to produce a visible image, or other devices. A digital data signal is applied to a computer or microprocessor, which interacts with a human user.

The radio waves from many transmitters pass through the air simultaneously without interfering with each other. They can be separated in the receiver because each transmitter's radio waves oscillate at a different rate, in other words each transmitter has a different frequency, measured in kilohertz (kHz), megahertz (MHz) or gigahertz (GHz). The bandpass filter in the receiver consists of a tuned circuit which acts like a resonator, similarly to a tuning fork. It has a natural resonant frequency at which it oscillates. The resonant frequency is set equal to the frequency of the desired radio station. The oscillating radio signal from the desired station causes the tuned circuit to oscillate in sympathy, and it passes the signal on to the rest of the receiver. Radio signals at other frequencies are blocked by the tuned circuit and not passed on.

Biological and Environmental Effects

Radio waves are *nonionizing radiation*, which means they do not have enough energy to separate electrons from atoms or molecules, ionizing them, or break chemical bonds, causing chemical reactions or DNA damage. The main effect of absorption of radio waves by materials is to heat them, similarly to the infrared waves radiated by sources of heat such as a space heater or wood fire. The oscillating electric field of the wave causes polar molecules to vibrate back and forth, increasing the temperature; this is how a microwave oven cooks food. However, unlike infrared waves, which are mainly absorbed at the surface of objects and cause surface heating, radio waves are able to penetrate the surface and deposit their energy inside materials and biological tissues. The depth to which radio waves penetrate decreases with their frequency, and also depends on the material's resistivity and permittivity; it is given by a parameter called the *skin depth* of the material, which is the depth within which 63% of the energy is deposited. For example, the 2.45 GHz radio waves (microwaves) in a microwave oven penetrate most foods approximately 2.5 to 3.8 cm (1 to 1.5 inches). Radio waves have been applied to the body for 100 years in the medical therapy of diathermy for deep heating of body tissue, to promote increased blood flow and healing. More recently they have been used to create higher temperatures in hyperthermia treatment and to kill cancer cells. Looking into a source of radio waves at close range, such as the waveguide of a working radio transmitter, can cause damage to the lens of the eye by heating. A strong enough beam of radio waves can penetrate the eye and heat the lens enough to cause cataracts.

Since the heating effect is in principle no different from other sources of heat, most research into possible health hazards of exposure to radio waves has focused on "nonthermal" effects; whether radio waves have any effect on tissues besides that caused by heating. Electromagnetic radiation has been classified by the International Agency for Research on Cancer (IARC) as "Possibly carcinogenic to humans". The conceivable evidence of cancer risk via Personal exposure to RF-EMF with mobile telephone use was identified.

Radio waves can be shielded against by a conductive metal sheet or screen, an enclosure of sheet or screen is called a Faraday cage. A metal screen shields against radio waves as well as a solid sheet as long as the holes in the screen are smaller than about 1/20 of wavelength of the waves.

Measurement

Since radio frequency radiation has both an electric and a magnetic component, it is often convenient to express intensity of radiation field in terms of units specific to each component. The unit *volts per meter* (V/m) is used for the electric component, and the unit *amperes per meter* (A/m) is used for the magnetic component. One can speak of an electromagnetic field, and these units are used to provide information about the levels of electric and magnetic field strength at a measurement location.

Another commonly used unit for characterizing an RF electromagnetic field is *power density*. Power density is most accurately used when the point of measurement is far enough away from the RF emitter to be located in what is referred to as the far field zone of the radiation pattern. In closer proximity to the transmitter, i.e., in the "near field" zone, the physical relationships between the electric and magnetic components of the field can be complex, and it is best to use the field strength units discussed above. Power density is measured in terms of power per unit area, for example, milliwatts per square centimeter (mW/cm²). When speaking of frequencies in the microwave range and higher, power density is usually used to express intensity since exposures that might occur would likely be in the far field zone.

RADIO PROPAGATION

Radio propagation is the behavior of radio waves as they travel, or are propagated, from one point to another, or into various parts of the atmosphere. As a form of electromagnetic radiation, like light waves, radio waves are affected by the phenomena of reflection, refraction, diffraction, absorption, polarization, and scattering. Understanding the effects of varying conditions on radio propagation has many practical applications, from choosing frequencies for international shortwave broadcasters, to designing reliable mobile telephone systems, to radio navigation, to operation of radar systems.

Several different types of propagation are used in practical radio transmission systems. Line-of-sight propagation means radio waves which travel in a straight line from the transmitting antenna to the receiving antenna. Line of sight transmission is used to medium range radio transmission such as cell phones, cordless phones, walkie-talkies, wireless networks, FM radio and television broadcasting and radar, and satellite communication, such as satellite television. Line-of-sight transmission on the surface of the Earth is limited to the distance to the visual horizon, which depends on the height of transmitting and receiving antennas. It is the only propagation method possible at microwave frequencies and above. At microwave frequencies, moisture in the atmosphere (rain fade) can degrade transmission.

At lower frequencies in the MF, LF, and VLF bands, due to diffraction radio waves can bend over obstacles like hills, and travel beyond the horizon as surface waves which follow the contour of the Earth. These are called ground waves. AM broadcasting stations use ground waves to cover their listening areas. As the frequency gets lower, the attenuation with distance decreases, so very low frequency (VLF) and extremely low frequency (ELF) ground waves can be used to communicate worldwide. VLF and ELF waves can penetrate significant distances through water and earth, and these frequencies are used for mine communication and military communication with submerged submarines.

At medium wave and shortwave frequencies (MF and HF bands) radio waves can refract from a layer of charged particles (ions) high in the atmosphere, called the ionosphere. This means that radio waves transmitted at an angle into the sky can be reflected back to Earth beyond the horizon, at great distances, even transcontinental distances. This is called skywave propagation. It is used by amateur radio operators to talk to other countries, and shortwave broadcasting stations that broadcast internationally. Skywave communication is variable, dependent on conditions in the upper atmosphere; it is most reliable at night and in the winter. Due to its unreliability, since the advent of communication satellites in the 1960s, many long range communication needs that previously used skywaves now use satellites.

In addition, there are several less common radio propagation mechanisms, such as tropospheric scattering (troposcatter) and near vertical incidence skywave (NVIS) which are used in specialized communication systems.

Free Space Propagation

In free space, all electromagnetic waves (radio, light, X-rays, etc.) obey the inverse-square law which states that the power density ρ of an electromagnetic wave is proportional to the inverse of the square of the distance r from a point source or:

$$\rho \propto \frac{1}{r^2}.$$

At typical communication distances from a transmitter, the transmitting antenna usually can be approximated by a point source. Doubling the distance of a receiver from a transmitter means that the power density of the radiated wave at that new location is reduced to one-quarter of its previous value.

The power density per surface unit is proportional to the product of the electric and magnetic field strengths. Thus, doubling the propagation path distance from the transmitter reduces each of these received field strengths over a free-space path by one-half.

Radio waves in vacuum travel at the speed of light. The Earth's atmosphere is thin enough that radio waves in the atmosphere travel very close to the speed of light, but variations in density and temperature can cause some slight refraction (bending) of waves over distances.

Modes

At different frequencies, radio waves travel through the atmosphere by different mechanisms or modes:

Radio frequencies and their primary mode of propagation				
	Band	Frequency	Wavelength	Propagation via
ELF	Extremely Low Frequency	3–30 Hz	100,000–10,000 km	Guided between the Earth and the D layer of the ionosphere.
SLF	Super Low Frequency	30–300 Hz	10,000–1,000 km	Guided between the Earth and the ionosphere.
ULF	Ultra Low Frequency	0.3–3 kHz (300–3,000 Hz)	1,000–100 km	Guided between the Earth and the ionosphere.
VLF	Very Low Frequency	3–30 kHz (3,000–30,000 Hz)	100–10 km	Guided between the Earth and the ionosphere.
LF	Low Frequency	30–300 kHz (30,000–300,000 Hz)	10–1 km	Guided between the Earth and the ionosphere. Ground waves.
MF	Medium Frequency	300–3000 kHz (300,000–3,000,000 Hz)	1000–100 m	Ground waves. E, F layer ionospheric refraction at night, when D layer absorption weakens.
HF	High Frequency (Short Wave)	3–30 MHz (3,000,000–30,000,000 Hz)	100–10 m	E layer ionospheric refraction. F1, F2 layer ionospheric refraction.
VHF	Very High Frequency	30–300 MHz (30,000,000–300,000,000 Hz)	10–1 m	Line-of-sight propagation. Infrequent E ionospheric (E_s) refraction. Uncommonly F2 layer ionospheric refraction during high sunspot activity up to 50 MHz and rarely to 80 MHz. Sometimes tropospheric ducting or meteor scatter
UHF	Ultra High Frequency	300–3000 MHz (300,000,000–3,000,000,000 Hz)	100–10 cm	Line-of-sight propagation. Sometimes tropospheric ducting.
SHF	Super High Frequency	3–30 GHz (3,000,000,000–30,000,000,000 Hz)	10–1 cm	Line-of-sight propagation. Sometimes rain scatter.
EHF	Extremely High Frequency	30–300 GHz (30,000,000,000–300,000,000,000 Hz)	10–1 mm	Line-of-sight propagation, limited by atmospheric absorption to a few kilometers
THF	Tremendously High frequency	0.3–3 THz (300,000,000,000–3,000,000,000,000 Hz)	1–0.1 mm	Line-of-sight propagation.

Direct Modes (Line-of-Sight)

Line-of-sight refers to radio waves which travel directly in a line from the transmitting antenna to the receiving antenna. It does not necessarily require a cleared sight path; at lower frequencies radio waves can pass through buildings, foliage and other obstructions. This is the most common propagation mode at VHF and above, and the only possible mode at microwave frequencies and above. On the surface of the Earth, line of sight propagation is limited by the visual horizon to about 40 miles (64 km). This is the method used by cell phones, cordless phones, walkie-talkies, wireless networks, point-to-point microwave radio relay links, FM and television broadcasting and radar. Satellite communication uses longer line-of-sight paths; for example home satellite dishes receive signals from communication satellites 22,000 miles (35,000 km) above the Earth, and ground stations can communicate with spacecraft billions of miles from Earth.

Ground plane reflection effects are an important factor in VHF line of sight propagation. The interference between the direct beam line-of-sight and the ground reflected beam often leads to an effective inverse-fourth-power (1/distance^4) law for ground-plane limited radiation.

Surface Modes (Groundwave)

Lower frequency (between 30 and 3,000 kHz) vertically polarized radio waves can travel as surface waves following the contour of the Earth; this is called *groundwave* propagation.

In this mode the radio wave propagates by interacting with the conductive surface of the Earth. The wave "clings" to the surface and thus follows the curvature of the Earth, so groundwaves can travel over mountains and beyond the horizon. Ground waves propagate in vertical polarization so vertical antennas (monopoles) are required. Since the ground is not a perfect electrical conductor, ground waves are attenuated as they follow the Earth's surface. Attenuation is proportional to frequency, so ground waves are the main mode of propagation at lower frequencies, in the MF, LF and VLF bands. Ground waves are used by radio broadcasting stations in the MF and LF bands, and for time signals and radio navigation systems.

At even lower frequencies, in the VLF to ELF bands, an Earth-ionosphere waveguide mechanism allows even longer range transmission. These frequencies are used for secure military communications. They can also penetrate to a significant depth into seawater, and so are used for one-way military communication to submerged submarines.

Early long distance radio communication (wireless telegraphy) before the mid-1920s used low frequencies in the longwave bands and relied exclusively on ground-wave propagation. Frequencies above 3 MHz were regarded as useless and were given to hobbyists (radio amateurs). The discovery around 1920 of the ionospheric reflection or skywave mechanism made the medium wave and short wave frequencies useful for long distance communication and they were allocated to commercial and military users.

Ionospheric Modes (Skywave)

Skywave propagation, also referred to as skip, is any of the modes that rely on reflection and refraction of radio waves from the ionosphere. The ionosphere is a region of the atmosphere from about 60 to 500 km (37 to 311 mi) that contains layers of charged particles (ions) which can refract a radio wave back toward the Earth. A radio wave directed at an angle into the sky can be reflected back to Earth beyond the horizon by these layers, allowing long distance radio transmission. The F2 layer is the most important ionospheric layer for long-distance, multiple-hop HF propagation, though F1, E, and D-layers also play significant roles. The D-layer, when present during sunlight periods, causes significant amount of signal loss, as does the E-layer whose maximum usable frequency can rise to 4 MHz and above and thus block higher frequency signals from reaching the F2-layer. The layers, or more appropriately "regions", are directly affected by the sun on a daily diurnal cycle, a seasonal cycle and the 11-year sunspot cycle and determine the utility of these modes. During solar maxima, or sunspot highs and peaks, the whole HF range up to 30 MHz can be used usually around the clock and F2 propagation up to 50 MHz is observed frequently depending upon daily solar flux 10.7cm radiation values. During solar minima, or minimum sunspot counts down to zero, propagation of frequencies above 15 MHz is generally unavailable.

Although the claim is commonly made that two-way HF propagation along a given path is reciprocal, that is, if the signal from location A reaches location B at a good strength, the signal from location B will be similar at station A because the same path is traversed in both directions. However, the ionosphere is far too complex and constantly changing to support the reciprocity theorem. The path is never exactly the same in both directions. In brief, conditions at the two terminii of a path generally cause dissimilar polarization shifts, dissimilar splits into ordinary rays and extraordinary or *Pedersen rays* which have difference propagation characteristics due to differences in ionization density, shifting zenith angles, effects of the Earth's magnetic dipole contours, antenna radiation patterns, ground conditions and other variables.

Forecasting of skywave modes is of considerable interest to amateur radio operators and commercial marine and aircraft communications, and also to shortwave broadcasters. Real-time propagation can be assessed by listening for transmissions from specific beacon transmitters.

Meteor Scattering

Meteor scattering relies on reflecting radio waves off the intensely ionized columns of air generated by meteors. While this mode is very short duration, often only from a fraction of second to couple of seconds per event, digital Meteor burst communications allows remote stations to communicate to a station that may be hundreds of miles up to over 1,000 miles (1,600 km) away, without the expense required for a satellite link. This mode is most generally useful on VHF frequencies between 30 and 250 MHz.

Auroral Backscatter

Intense columns of Auroral ionization at 100 km altitudes within the auroral oval backscatter radio waves, including those on HF and VHF. Backscatter is angle-sensitive—incident ray vs. magnetic field line of the column must be very close to right-angle. Random motions of electrons spiraling around the field lines create a Doppler-spread that broadens the spectra of the emission to more or less noise-like—depending on how high radio frequency is used. The radio-auroras are observed mostly at high latitudes and rarely extend down to middle latitudes. The occurrence of radio-auroras depends on solar activity (flares, coronal holes, CMEs) and annually the events are more numerous during solar cycle maxima. Radio aurora includes the so-called afternoon radio aurora which produces stronger but more distorted signals and after the Harang-minima, the late-night radio aurora (sub-storming phase) returns with variable signal strength and lesser doppler spread. The propagation range for this predominantly back-scatter mode extends up to about 2000 km in east-west plane, but strongest signals are observed most frequently from the north at nearby sites on same latitudes.

Rarely, a strong radio-aurora is followed by Auroral-E, which resembles both propagation types in some ways.

Sporadic-E Propagation

Sporadic E (Es) propagation can be observed on HF and VHF bands. It must not be confused with ordinary HF E-layer propagation. Sporadic-E at mid-latitudes occurs mostly during summer season, from May to August in the northern hemisphere and from November to February in the southern hemisphere. There is no single cause for this mysterious propagation mode. The reflection takes place in a thin sheet of ionisation around 90 km height. The ionisation patches drift westwards at speeds of few hundred km per hour. There is a weak periodicity noted during the season and typically Es is observed on 1 to 3 successive days and remains absent for a few days to reoccur again. Es do not occur during small hours; the events usually begin at dawn, and there is a peak in the afternoon and a second peak in the evening. Es propagation is usually gone by local midnight.

Observation of radio propagation beacons operating around 28.2 MHz, 50 MHz and 70 MHz, indicates that maximum observed frequency (MOF) for Es is found to be lurking around 30 MHz on most days during the summer season, but sometimes MOF may shoot up to 100 MHz or even more in ten minutes to decline slowly during the next few hours. The peak-phase includes oscillation of MOF with periodicity of approximately 5-10 minutes. The propagation range for Es single-hop is typically 1000 to 2000 km, but with multi-hop, double range is observed. The signals are very strong but also with slow deep fading.

Tropospheric Modes

Radio waves in the VHF and UHF bands can travel somewhat beyond the visual horizon due to refraction in the troposphere, the bottom layer of the atmosphere below 20 km. This is due to changes in the refractive index of air with temperature and pressure. Tropospheric delay is a source of error in radio ranging techniques, such as the Global Positioning System (GPS). In addition, unusual conditions can sometimes allow propagation at greater distances:

Tropospheric Ducting

Sudden changes in the atmosphere's vertical moisture content and temperature profiles can on random occasions make UHF, VHF and microwave signals propagate hundreds of kilometers up to about 2,000 kilometers (1,200 miles)—and for ducting mode even farther—beyond the normal radio-horizon. The inversion layer is mostly observed over high pressure regions, but there are several tropospheric weather conditions which create these randomly occurring propagation modes. Inversion layer's altitude for non-ducting is typically found between 100 and 1,000 meters (330 and 3,280 feet) and for ducting about 500 to 3,000 meters (1,600 to 9,800 feet), and the duration of the events are typically from several hours up to several days. Higher frequencies experience the most dramatic increase of signal strengths, while on low-VHF and HF the effect is negligible. Propagation path attenuation may be below free-space loss. Some of the lesser inversion types related to warm ground and cooler air moisture content occur regularly at certain times of the year and time of day. A typical example could be the late summer, early morning tropospheric enhancements that bring in signals from distances up to few hundred kilometers for a couple of hours, until undone by the Sun's warming effect.

Tropospheric Scattering (Troposcatter)

At VHF and higher frequencies, small variations (turbulence) in the density of the atmosphere at a height of around 6 miles (9.7 km) can scatter some of the normally line-of-sight beam of radio frequency energy back toward the ground. In tropospheric scatter (troposcatter) communication systems a powerful beam of microwaves is aimed above the horizon, and a high gain antenna over the horizon aimed at the section of the troposphere though which the beam passes receives the tiny scattered signal. Troposcatter systems can achieve over-the-horizon communication between stations 500 miles (800 km) apart, and the military developed networks such as the White Alice Communications System covering all of Alaska before the 1960s, when communication satellites largely replaced them.

Rain Scattering

Rain scattering is purely a microwave propagation mode and is best observed around 10 GHz, but extends down to a few gigahertz—the limit being the size of the scattering particle size vs. wavelength. This mode scatters signals mostly forwards and backwards when using horizontal polarization and side-scattering with vertical polarization. Forward-scattering typically yields propagation ranges of 800 km. Scattering from snowflakes and ice pellets also occurs, but scattering from ice without watery surface is less effective. The most common application for this phenomenon is microwave rain radar, but rain scatter propagation can be a nuisance causing unwanted signals to intermittently propagate where they are not anticipated or desired. Similar reflections may also occur from insects though at lower altitudes and shorter range. Rain also causes attenuation of point-to-point and satellite microwave links. Attenuation values up to 30 dB have been observed on 30 GHz during heavy tropical rain.

Airplane Scattering

Airplane scattering (or most often reflection) is observed on VHF through microwaves and, besides back-scattering, yields momentary propagation up to 500 km even in mountainous terrain. The most common back-scatter applications are air-traffic radar, bistatic forward-scatter guided-missile and airplane-detecting trip-wire radar, and the US space radar.

Lightning Scattering

Lightning scattering has sometimes been observed on VHF and UHF over distances of about 500 km. The hot lightning channel scatters radio-waves for a fraction of a second. The RF noise burst from the lightning makes the initial part of the open channel unusable and the ionization disappears quickly because of recombination at low altitude and high atmospheric pressure. Although the hot lightning channel is briefly observable with microwave radar, no practical use for this mode has been found in communications.

Other Effects

Diffraction

Knife-edge diffraction is the propagation mode where radio waves are bent around sharp edges. For example, this mode is used to send radio signals over a mountain range when a line-of-sight path is not available. However, the angle cannot be too sharp or the signal will not diffract. The diffraction mode requires increased signal strength, so higher power or better antennas will be needed than for an equivalent line-of-sight path.

Diffraction depends on the relationship between the wavelength and the size of the obstacle. In other words, the size of the obstacle in wavelengths. Lower frequencies diffract around large smooth obstacles such as hills more easily. For example, in many cases where VHF (or higher frequency) communication is not possible due to shadowing by a hill, it is still possible to communicate using the upper part of the HF band where the surface wave is of little use.

Diffraction phenomena by small obstacles are also important at high frequencies. Signals for urban cellular telephony tend to be dominated by ground-plane effects as they travel over the rooftops of the urban environment. They then diffract over roof edges into the street, where multipath propagation, absorption and diffraction phenomena dominate.

Absorption

Low-frequency radio waves travel easily through brick and stone and VLF even penetrates sea-water. As the frequency rises, absorption effects become more important. At microwave or higher frequencies, absorption by molecular resonances in the atmosphere (mostly from water, H_2O and oxygen, O_2) is a major factor in radio propagation. For example, in the 58–60 GHz band, there is a major absorption peak which makes this band useless for long-distance use. This phenomenon was first discovered during radar research in World War II. Above about 400 GHz, the Earth's atmosphere blocks most of the spectrum while still passing some - up to UV light, which is blocked by ozone - but visible light and some of the near-infrared is transmitted. Heavy rain and falling snow also affect microwave absorption.

Measuring HF propagation

HF propagation conditions can be simulated using radio propagation models, such as the Voice of America Coverage Analysis Program, and realtime measurements can be done using chirp transmitters. For radio amateurs the WSPR mode provides maps with real time propagation conditions between a network of transmitters and receivers. Even without special beacons the realtime propagation conditions can be measured: a worldwide network of receivers decodes morse code signals on amateur radio frequencies in realtime and provides sophisticated search functions and propagation maps for every station received.

Practical Effects

The average person can notice the effects of changes in radio propagation in several ways.

In AM broadcasting, the dramatic ionospheric changes that occur overnight in the mediumwave band drive a unique broadcast license scheme, with entirely different transmitter power output levels and directional antenna patterns to cope with skywave propagation at night. Very few stations are allowed to run without modifications during dark hours, typically only those on clear channels in North America. Many stations have no authorization to run at all outside of daylight hours. Otherwise, there would be nothing but interference on the entire broadcast band from dusk until dawn without these modifications.

For FM broadcasting (and the few remaining low-band TV stations), weather is the primary cause for changes in VHF propagation, along with some diurnal changes when the sky is mostly without cloud cover. These changes are most obvious during temperature inversions, such as in the late-night and early-morning hours when it is clear, allowing the ground and the air near it to cool more rapidly. This not only causes dew, frost, or fog, but also causes a slight "drag" on the bottom of the radio waves, bending the signals down such that they can follow the Earth's curvature over the normal radio horizon. The result is typically several stations being heard from another media market — usually a neighboring one, but sometimes ones from a few hundred kilometers away. Ice storms are also the result of inversions, but these normally cause more scattered omnidirection propagation, resulting mainly in interference, often among weather radio stations. In late spring and early summer, a combination of other atmospheric factors can occasionally cause skips that duct high-power signals to places well over 1000 km away.

Non-broadcast signals are also affected. Mobile phone signals are in the UHF band, ranging from 700 to over 2600 Megahertz, a range which makes them even more prone to weather-induced propagation changes. In urban (and to some extent suburban) areas with a high population density, this is partly offset by the use of smaller cells, which use lower effective radiated power and beam tilt to reduce interference, and therefore increase frequency reuse and user capacity. However, since this would not be very cost-effective in more rural areas, these cells are larger and so more likely to cause interference over longer distances when propagation conditions allow.

While this is generally transparent to the user thanks to the way that cellular networks handle cell-to-cell handoffs, when cross-border signals are involved, unexpected charges for international roaming may occur despite not having left the country at all. This often occurs between southern San Diego and northern Tijuana at the western end of the U.S./Mexico border, and between eastern Detroit and western Windsor along the U.S./Canada border. Since signals can travel unobstructed over a body of water far larger than the Detroit River, and cool water temperatures also cause inversions in surface air, this "fringe roaming" sometimes occurs across the Great Lakes, and between islands in the Caribbean. Signals can skip from the Dominican Republic to a mountainside in Puerto Rico and vice versa, or between the U.S. and British Virgin Islands, among others. While unintended cross-border roaming is often automatically removed by mobile phone company billing systems, inter-island roaming is typically not.

GROUND WAVE PROPAGATION

Ground Wave propagation is a method of radio wave propagation that uses the area between the surface of the earth and the ionosphere for transmission. The ground wave can propagate a considerable distance over the earth's surface particularly in the low frequency and medium frequency portion of the radio spectrum.

Ground wave radio signal propagation is ideal for relatively short distance propagation on these frequencies during the daytime. Sky-wave ionospheric propagation is not possible during the day because of the attenuation of the signals on these frequencies caused by the D region in the ionosphere. In view of this, lower frequency radio communications stations need to rely on the ground-wave propagation to achieve their coverage.

Typically, what is referred to as a *ground wave* radio signal is made up of a number of constituent waves. If the antennas are in the line of sight then there will be a direct wave as well as a reflected signal. As the names suggest the direct signal is one that travels directly between the two antennas and is not affected by the locality. There will also be a reflected signal as the transmission will be reflected by a number of objects including the earth's surface and any hills, or large buildings that may be present. In addition to this there is a surface wave. This tends to follow the curvature of the Earth and enables coverage beyond the horizon. It is the sum of all these components that is known as the ground wave. Beyond the horizon the direct and reflected waves are blocked by the curvature of the Earth, and the signal is purely made up of the diffracted surface wave. It is for this reason that surface wave is commonly called ground wave propagation.

Surface Wave

The radio signal spreads out from the transmitter along the surface of the Earth. Instead of just travelling in a straight line the radio signals tend to follow the curvature of the Earth. This is because currents are induced in the surface of the earth and this action slows down the wave-front in this region, causing the wave-front of the radio communications signal to tilt downwards towards the Earth. With the wave-front tilted in this direction it is able to curve around the Earth and be received well beyond the horizon.

Effect of Frequency on Ground Wave Propagation

As the wavefront of the ground wave travels along the Earth's surface it is attenuated. The degree of attenuation is dependent upon a variety of factors. Frequency of the radio signal is one of the major determining factor as losses rise with increasing frequency. As a result, it makes this form of propagation impracticable above the bottom end of the HF portion of the spectrum (3 MHz). Typically a signal at 3.0 MHz will suffer an attenuation that may be in the region of 20 to 60 dB more than one at 0.5 MHz dependent upon a variety of factors in the signal path including the distance. In view of this it can be seen why even high power HF radio broadcast stations may only be audible for a few miles from the transmitting site via the ground wave.

Effect of the Ground

The surface wave is also very dependent upon the nature of the ground over which the signal travels. Ground conductivity, terrain roughness and the dielectric constant all affect the signal attenuation. In addition to this the ground penetration varies, becoming greater at lower frequencies, and this means that it is not just the surface conductivity that is of interest. At the higher frequencies this is not of great importance, but at lower frequencies penetration means that ground strata down to 100 meters may have an effect.

Despite all these variables, it is found that terrain with good conductivity gives the best result. Thus soil type and the moisture content are of importance. Salty sea water is the best, and rich agricultural, or marshy land is also good. Dry sandy terrain and city centres are by far the worst. This means sea paths are optimum, although even these are subject to variations due to the roughness of the sea, resulting on path losses being slightly dependent upon the weather. It should also be noted that in view of the fact that signal penetration has an effect, the water table may have an effect dependent upon the frequency in use.

Polarization and Ground Wave Propagation

The type of antenna and its polarization has a major effect on ground wave propagation. Vertical polarization is subject to considerably less attenuation than horizontally polarized signals. In some cases the difference can amount to several tens of decibels. It is for this reason that medium wave broadcast stations use vertical antennas, even if they have to be made physically short by adding inductive loading. Ships making use of the MF marine bands often use inverted L antennas as these are able to radiate a significant proportion of the signal that is vertically polarized.

At distances that are typically towards the edge of the ground wave coverage area, some sky-wave signal may also be present, especially at night when the D layer attenuation is reduced. This may serve to reinforce or cancel the overall signal resulting in figures that will differ from those that may be expected.

ANOMALOUS PROPAGATION

Anomalous propagation (sometimes shortened to anaprop or anoprop) includes different forms of radio propagation due to an unusual distribution of temperature and humidity with height in the atmosphere. While this includes propagation with larger losses than in a standard atmosphere, in practical applications it is most often meant to refer to cases when signal propagates beyond normal radio horizon.

Anomalous propagation can cause interference to VHF and UHF radio communications if distant stations are using the same frequency as local services. Over-the-air analog television broadcasting, for example, may be disrupted by distant stations on the same channel, or experience distortion of transmitted signals ghosting). Radar systems may produce inaccurate ranges or bearings to distant targets if the radar "beam" is bent by propagation effects. However, radio hobbyists take advantage of these effects in TV and FM DX.

Causes

Air Temperature Profile

The first assumption of the prediction of propagation of a radio wave is that it is moving through air with temperature that declines at a standard rate with height in the troposphere. This has the effect of slightly bending (refracting) the path toward the Earth, and accounts for an effective range that is slightly greater than the geometric distance to the horizon. Any variation to this stratification of temperatures will modify the path followed by the wave.

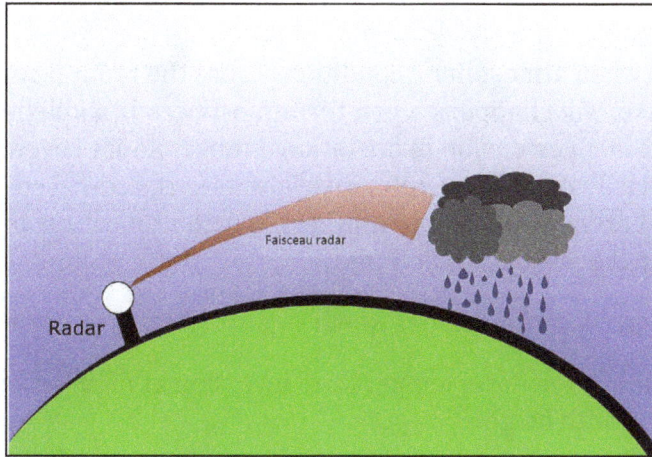

Super refraction in radar operation.

Sub-refraction

Sub-refraction occurs when atmospheric conditions cause the radar beam to bend less than in the Standard Atmosphere. This happens when the atmosphere is unstable relative to the Standard Atmosphere. Sub-refraction causes the radar to overshoot targets that would ordinarily be observed under standard atmospheric conditions (Figure 1) and results in a reduction in the operational range of the radar. In addition to underestimated echo heights, this phenomenon tends to reduce ground clutter in the lowest elevation cuts.

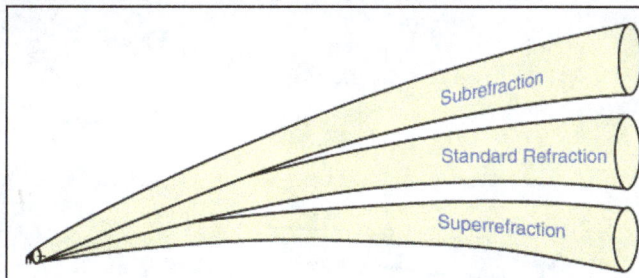

Sub- and super-refraction in comparison with standard refraction.

The classic situation under which subrefraction occurs is the inverted-vee sounding, characterized by a dry-adiabatic or superadiabatic temperature lapse rate and a constant or increasing moisture supply with height. The near-surface air is typically dry. This situation is common over desert areas and on the lee side of mountain ranges, especially during the afternoon. In order to detect the tops, use of a lower elevation slice would be required, which would result in the actual tops correspondingly being displayed at a lower height than the true altitude.

Besides overestimated echo heights, super-refraction increases ground clutter in the lowest elevation cuts and is the cause of what we normally refer to as anomalous propagation or AP echoes that are encountered if the density/height profile is significantly different from normal, and the beam will be bent much more, or much less, than normal. In order to detect the tops, use of a higher elevation slice would be required, which would result in the tops being displayed at a correspondingly higher height than their true altitude.

Super-refraction

Super-refraction occurs when atmospheric conditions cause the radar beam to bend more than in the Standard Atmosphere. This happens when the atmosphere is stable relative to the Standard Atmosphere, and results in an extension in operational range. Radar coverage can be extended up to 150% of normal, especially if the super-refractive layer is surfaced-based. If the beam is not bent enough to intersect the Earth's surface, low altitude precipitation echoes that would ordinarily be below a standard refracted beam can be detected.

Several common situations/locations under which superrefraction or beam trapping occurs are:

1. The development of a low-level temperature inversion at night associated with a sharp decrease of moisture with height.

2. A flow of warm, dry air advects over a cooler surface; this is most effective if the cooler surface is water – due to mixing, lower layers of the atmosphere are cooled and moistened.

3. The development of a sea breeze with cool moist air moving inland beneath a warm dry continental air mass.

4. Rain-cooled outflow from a thunderstorm forms a surface-based temperature inversion.

5. Location of the tropopause, which denotes an area of significant decrease of lapse rate with height.

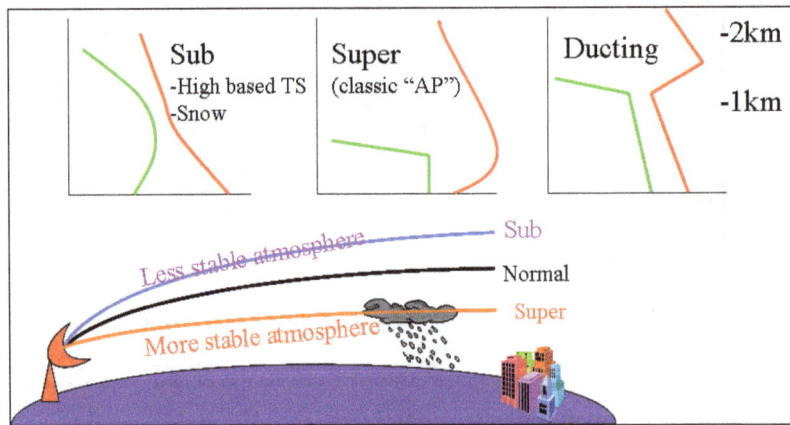

Typical sounding profiles sub- and superrefraction and ducting.

Super-refraction results in overestimates of heights measured by radar. When super-refraction is occurring, a precipitation target is observed at a higher elevation angle than standard. The beam is closer to the ground than the standard atmosphere-based charts indicate, so the antenna must be raised to a greater elevation angle than normal to find the top of an echo.

Effects of super-refraction have a greater impact on the radar operator than do the effects of sub-refraction. Since the beam is lower than standard, low-altitude targets that would ordinarily be below the beam can be detected, as was previously stated.

Unfortunately, super-refraction can result in more range-folded echoes being detected. These echoes occur when super-refraction bends the radar beam so that it tends to follow the Earth's curvature for long distances and targets beyond the unambiguous range of the radar are detected.

Another unfavourable effect of super-refraction concerns the ability of the radar to detect non-precipitation targets at extended ranges. Ground clutter, a highly reflective echo pattern typically generated from terrain features and other objects close to the radar, will be expanded under super-refractive conditions. Surface features will be detected and displayed at extended ranges when the radar beam is bent enough to travel near the ground, or bounce along the ground, for a long distance. The expanded ground clutter pattern is typically referred to as anomalous propagation (AP) by radar operators to distinguish it from precipitation echoes. The usefulness of the radar is diminished because it is often difficult to distinguish ground return from precipitation targets.

Echo properties which can be used to identify AP-induced echoes on conventional weather radars include:

1. A persistent pattern for a given elevation angle, but a rapidly changing patter with change of elevation angle, and

2. A large number of small-scale features, particularly high-intensity reflectivity cores with abnormally large reflectivity gradients (greater than 20 dB change over a distance of 1 km).

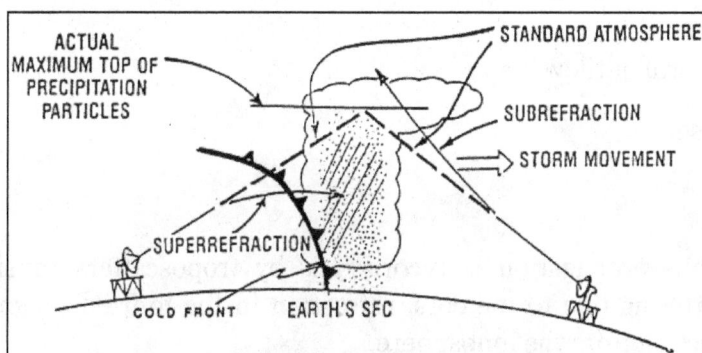

Refraction of the radar beam with associated atmospheric conditions and consequences for the returned signal

In addition to pattern recognition, the radar operator will identify AP-induced echoes with more skill if he/she is aware of current atmospheric conditions. A radar operator can actually monitor the atmospheric stability by noting how the ground clutter pattern changes for a fixed elevation angle over time. Typically, ground clutter maximizes during the late night/early morning (radiation inversion – surface-based stable layer) and minimizes during the late afternoon/early evening (dry adiabatic lapse rate – surface-based unstable layer).

Ducting

Ducting is a special superrefractive condition such that the radar "beam" gets trapped or "ducted" within a stable layer or temperature inversion. This causes the beam to bend downward more than normal, but the beam rarely comes in contact with the ground and little energy is lost through attenuation. Operationally, this is an extreme case of superrefraction which can result in the detection of targets well beyond the operating Rmax.

Ducting properties:

- Beam bends towards earth and bounces within trapping layer.

- Very large increase in effective range.

- Situations – strong inversion, with sharp decrease of moisture with height:

  ◦ Nocturnal inversion.

  ◦ Warm air flow over cool water.

  ◦ Sea breeze.

  ◦ Thunderstorm outflow.

  ◦ Tropopause.

Other Causes

Other ways anomalous propagation is recorded is by troposcatters causing irregularities in the troposphere, scattering due to meteors, refraction in the ionized regions and layers of the ionosphere, and reflection from the ionosphere.

Finally, multipath propagation near the Earth's surface has multiple causes, including atmospheric ducting, ionospheric reflection and refraction, and reflection from water bodies and terrestrial objects such as mountains and buildings.

In Radio

Anomalous propagation can be a limiting factor for the propagation of radiowaves, especially the super refraction. However, reflection on the ionosphere is a common use of this phenomenon to extend the range of the signal. Other multiple reflections or refractions are more complex to predict but can be still useful.

Radar

The position of the radar echoes depend heavily on the standard decrease of temperature hypothesis. However, the real atmosphere can vary greatly from the norm. *Anomalous Propagation* (AP) refers to false radar echoes usually observed when calm, stable atmospheric conditions, often associated with super refraction in a temperature inversion, direct the radar beam toward the ground. The processing program will then wrongly place the return echoes at the height and distance it would have been in normal conditions.

This type of false return is relatively easy to spot on a time loop if it is due to night cooling or marine inversion as one sees very strong echoes developing over an area, spreading in size laterally, not moving but varying greatly in intensity with time. After sunrise, the inversion disappears gradually and the area diminishes correspondingly. Inversion of temperature exists too ahead of warm fronts, and around thunderstorms' cold pool. Since precipitation exists in those circumstances, the abnormal propagation echoes are then mixed with real rain and targets of interest, which make them more difficult to separate.

Anomalous Propagation is different from *ground clutter*, ocean reflections (*sea clutter*), biological returns from birds and insects, debris, chaff, sand storms, volcanic eruption plumes, and other non-precipitation meteorological phenomena. Ground and sea clutters are permanent reflection from fixed areas on the surface with stable reflective characteristics. Biological scatterer gives weak echoes over a large surface. These can vary in size with time but not much in intensity. Debris and chaff are transient and move in height with time. They are all indicating something actually there and either relevant to the radar operator and readily explicable and theoretically able to be reproduced. AP in the sense of radar is colloquially known as "garbish" and ground clutter as "rubbage".

Doppler radars and Pulse-Doppler radars are extracting the velocities of the targets. Since AP comes from stable targets, it is possible to subtract the reflectivity data having a null speed and clean the radar images. Ground, sea clutter and the energy spike from the sun setting can be distinguished the same way but not other artifacts. This method is used in most modern radars, including air traffic control and weather radars.

SKY WAVE PROPAGATION

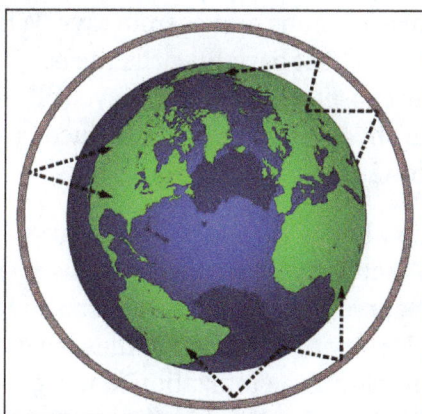

Radio waves (black) reflecting off the ionosphere (red) during skywave propagation

In radio communication, skywave or skip refers to the propagation of radio waves reflected or refracted back toward Earth from the ionosphere, an electrically charged layer of the upper atmosphere. Since it is not limited by the curvature of the Earth, skywave propagation can be used to communicate beyond the horizon, at intercontinental distances. It is mostly used in the short-wave frequency bands.

As a result of skywave propagation, a signal from a distant AM broadcasting station, a shortwave station, or – during sporadic E propagation conditions (principally during the summer months in both hemispheres) a distant VHF FM or TV station – can sometimes be received as clearly as local stations. Most long-distance shortwave (high frequency) radio communication – between 3 and 30 MHz – is a result of skywave propagation. Since the early 1920s amateur radio operators (or "hams"), limited to lower transmitter power than broadcast stations, have taken advantage of skywave for long distance (or "DX") communication.

Skywave propagation is distinct from:

- Tropospheric scatter, an alternative method of achieving over-the-horizon transmission at higher frequencies,

- Groundwave propagation, where radio waves travel along Earth's surface without being reflected or refracted by the atmosphere – the dominant propagation mode at lower frequencies,

- Line-of-sight propagation, in which radio waves travel in a straight line, the dominant mode at higher frequencies.

Local and Distant Skywave Propagation

Skywave transmissions can be used for long distance communications (DX) by waves directed at a low angle as well as relatively local communications via nearly-vertically directed waves (Near Vertical Incidence Skywaves – NVIS).

Low Angle Skywaves

The ionosphere is a region of the upper atmosphere, from about 80 km to 1000 km in altitude, where neutral air is ionized by solar photons and cosmic rays. When high frequency signals enter the ionosphere at a low angle they are bent back towards the earth by the ionized layer. If the peak ionization is strong enough for the chosen frequency, a wave will exit the bottom of the layer earthwards – as if obliquely reflected from a mirror. Earth's surface (ground or water) then reflects the descending wave back up again towards the ionosphere.

When operating at frequencies just below the MUF losses can be quite small, so the radio signal may effectively "bounce" or "skip" between the earth and ionosphere two or more times (multi-hop propagation), even following the curvature of the earth. Consequently, even signals of only a few Watts can sometimes be received many thousands of miles away. This is what enables shortwave broadcasts to travel all over the world. If the ionization is not great enough, the wave only curves slightly downwards, and subsequently upwards as the ionization peak is passed so that it exits the top of the layer only slightly displaced. The wave then is lost in space. To prevent this a lower frequency must be chosen. With a single "hop", path distances up to 3500 km may be reached. Longer transmissions can occur with two or more hops.

Near-vertical Skywaves

Skywaves directed vertically, or almost vertically, are called NVIS for "Near-Vertical Incidence". At some frequencies, generally in the lower shortwave region, the high angle skywaves will be reflected directly back towards the ground. When the wave returns to ground it is spread out over a wide area, allowing communications within several hundred miles of the transmitting antenna. NVIS enables local plus regional communications, even from low-lying valleys, to a large area, for example, an entire state or small country. Coverage of a similar area via a line-of-sight VHF transmitter would require a very high mountain top location. NVIS is thus useful for state wide networks as might be needed for emergency communications. In short wave broadcasting, NVIS is very useful for regional broadcasts that are targeted to an area that extends out from the transmitter location to a few hundred miles,such as would be the case in a country or language group to be reached from within the borders of that country. This will be much more economical that using multiple FM (VHF) or AM broadcast transmitters. Suitable antennas are designed to produce a strong lobe at high angles. When short range skywave is undesirable, as when an AM broadcaster wishes to avoid interference between the ground wave and sky wave, Anti-fading antennas are used to suppress the waves being propagated at the higher angles.

Intermediate Distance Coverage

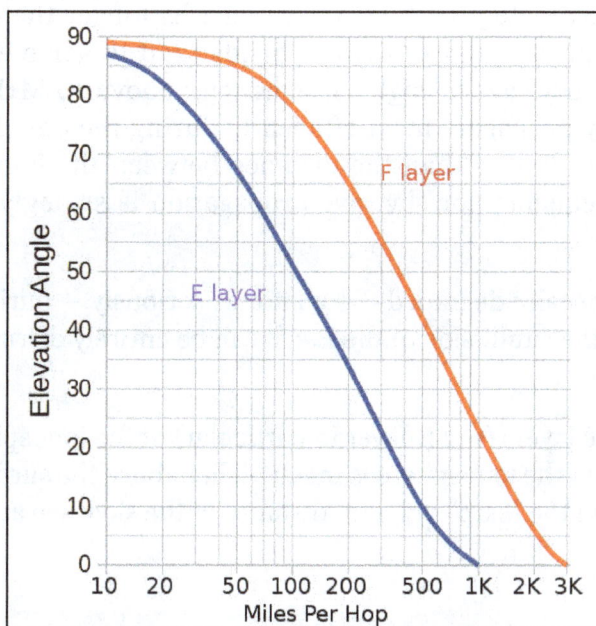

Antenna vertical angle required vs distance for skywave propagation

For every distance, from local to maximum distance transmission, (DX), there is an optimum "take off" angle for the antenna, as shown here. For example, using the F layer during the night, to best reach a receiver 500 miles away, an antenna should be chosen that has a strong lobe at 40 degrees elevation. One can also see that for the longest distances, a lobe at low angles (below 10 degrees) is best. For NVIS, angles above 45 degrees are optimum. Suitable antennas for long distance would be a high Yagi or a rhombic, for NVIS would be a dipole or array of dipoles about 0.2 wavelengths above ground, and for intermediate distances would be a dipole or Yagi at about 0.5 wavelengths above ground. Vertical patterns for each type of antenna are used to select the proper antenna.

Fading

At any distance sky waves will fade. The layer of ionospheric plasma with sufficient ionization (the reflective surface) is not fixed, but undulates like the surface of the ocean. Varying reflection efficiency from this changing surface can cause the reflected signal strength to change, causing *"fading"* in shortwave broadcasts. Even more serious fading can occur when signals arrive via two or more paths, for example when both single hop and double hop waves interfere with other, or when a skywave signal and a ground wave signal arrive at about the same strength. This is the most common source of fading with night time AM broadcast signals. Fading is always present with sky wave signals, and except for digital signals such as DRM seriously limit the fidelity of short wave broadcasts.

Other Considerations

VHF signals with frequencies above about 30 MHz usually penetrate the ionosphere and are not returned to the Earth's surface. E-skip is a notable exception, where VHF signals including FM broadcast and VHF TV signals are frequently reflected to the Earth during late spring and early summer. E-skip rarely affects UHF frequencies, except for very rare occurrences below 500 MHz.

Frequencies below approximately 10 MHz (wave lengths longer than 30 meters), including broadcasts in the mediumwave and shortwave bands (and to some extent longwave), propagate most efficiently by skywave at night. Frequencies above 10 MHz (wavelengths shorter than 30 meters) typically propagate most efficiently during the day. Frequencies lower than 3 kHz have a wave length longer than the distance between the Earth and the ionosphere. The maximum usable frequency for skywave propagation is strongly influenced by sunspot number.

Skywave propagation is usually degraded – sometimes seriously – during geomagnetic storms. Skywave propagation on the sunlit side of the Earth can be entirely disrupted during sudden ionospheric disturbances.

Because the lower-altitude layers (the E-layer in particular) of the ionosphere largely disappear at night, the refractive layer of the ionosphere is much higher above the surface of the Earth at night. This leads to an increase in the "skip" or "hop" distance of the skywave at night.

SPACE WAVE PROPAGATION

The radio waves having high frequencies are basically called as space waves. These waves have the ability to propagate through atmosphere, from transmitter antenna to receiver antenna. These waves can travel directly or can travel after reflecting from earth's surface to the troposphere surface of earth. So, it is also called as Tropospherical Propagation. In the diagram of medium wave propagation, c shows the space wave propagation. Basically the technique of space wave propagation is used in bands having very high frequencies. E.g. V.H.F. band, U.H.F band etc. At such higher frequencies the other wave propagation techniques like sky wave propagation, ground wave propagation can't work. Only space wave propagation is left which can handle frequency waves of higher frequencies. The other name of space wave propagation is line of sight propagation. There are some limitations of space wave propagation.

1. These waves are limited to the curvature of the earth.

2. These waves have line of sight propagation, means their propagation is along the line of sight distance.

The line of sight distance is that exact distance at which both the sender and receiver antenna are in sight of each other. So, from the above line it is clear that if we want to increase the transmission distance then this can be done by simply extending the heights of both the sender as well as the receiver antenna. This type of propagation is used basically in radar and television communication.

The frequency range for television signals is nearly 80 to 200MHz. These waves are not reflected by the ionosphere of the earth. The property of following the earth's curvature is also missing in these waves. So, for the propagation of television signal, geostationary satellites are used. The satellites complete the task of reflecting television signals towards earth. If we need greater transmission then we have to build extremely tall antennas.

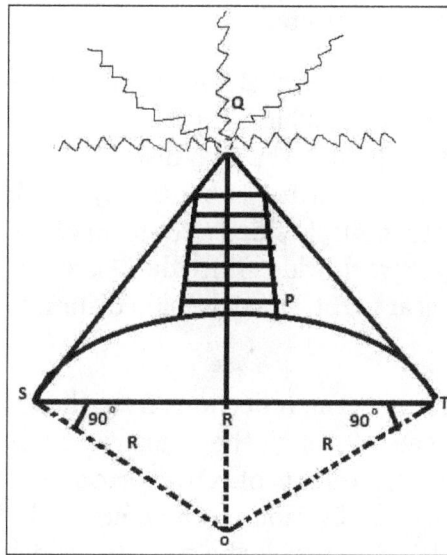

IONOSPHERIC PROPAGATION

The ionosphere exists between about 90 and 1000 km above the earth's surface. Radiation from the sun ionizes atoms and molecules here, liberating electrons from molecules and creating a space of free electron and ions. Subjected to an external electric field from a radio signal, these free and ions will experience a force and be pushed into motion. However, since the mass of the ions is much larger than the mass of the electrons, ionic motions are relatively small and will be ignored here.

Free electron densities on the order of 10^{10} to 10^{12} electrons per cubic metre are produced by ionization from the sun's rays. Layers of high densities of electrons are given special names called the D, E, and F layers, as shown in figure. During the day the F layer splits into two layers called the F_1 and F_2 layers, while the D layer vanishes completely at night.

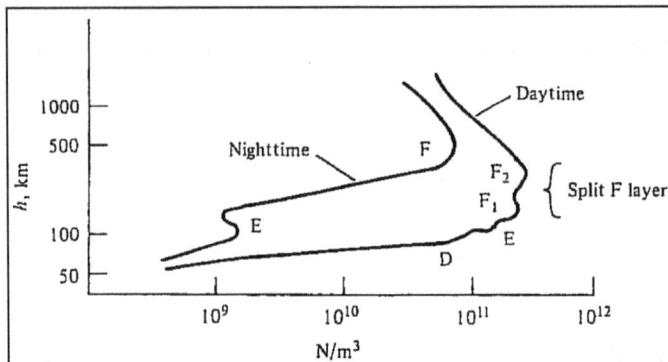

Electron density as a function of altitude, and various ionospheric layers

Radio waves below 40 MHz are significantly affected by the ionosphere, primarily because radio waves in this frequency range are effectively reflected by the ionosphere. The E and F layers are the most important for this process. For frequencies beyond 40 MHz, the wave tend to penetrate through the atmosphere versus being reflected.

The major usefulness of the ionosphere is that the reflections enable wave propagation over a much larger distance than would be possible with line-of-sight or even atmospheric refraction effects. This is shown graphically in figure. The skip distance d_{max} can be very large, allowing very large communication distances. This is further enhanced by multiple reflections between the ionosphere and the ground, leading to multiple skips. This form of propagation allows shortwave and amateur radio signals to propagate worldwide. Since the D layer disappears at night, the best time for long-range communications is at night, since the skip distance is larger as the E, and F regions are at higher altitudes.

Where does the reflection come from? The reflections from the ionosphere are actually produced by refraction as the wave propagates through the ionosphere. The ionosphere is a concentrated region highly charged ions and electrons that collective form an ionized gas or plasma. This gas has a dielectric constant that is a function of various parameters, including the electron concentration and the frequency of operation. We now derive the dielectric constant of a plasma.

The electric field will produce a force on a given electron and displace it along a vector $\vec{r}$, as shown in the figure.

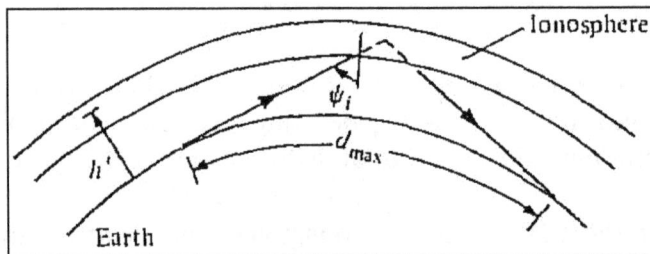

A single skip of a radio wave using the ionosphere.

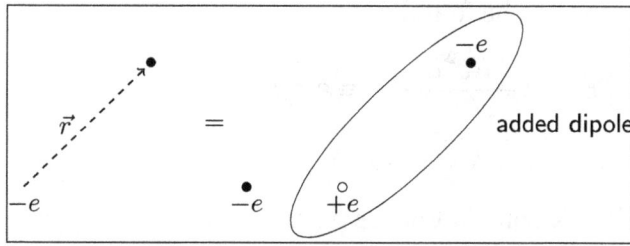

Representation of a moving electron as a dipole moment.

The displacement of an electron along this path can be modelled in an equivalent situation where the original electron remained stationary and an equivalent electric dipole is added, as shown in the right half of the figure. The dipole moment of this dipole is then equal to $-e\vec{r}$. If there are N electrons per unit volume, each displaced by $\vec{r}$ on average, then the volume polarization is,

$$\vec{P} = -Ne\vec{r}.$$

The equation of motion for a single electron of mass $m_e = 9.109 \times 10^{-31}$ kg, charge $e = 1.6021 \times 10 - 19C$, with velocity $\vec{v} = d\vec{r} / dt$, and acted upon by an electric field $\vec{\varepsilon}$, is,

$$m_e \frac{d\vec{v}}{dt} = -e\vec{\varepsilon}$$

The electron also experiences a frictional force resulting from collisions with neutral molecules. This force is added to the electric field force above, yielding,

$$m_e \frac{d\vec{v}}{dt} = -e\vec{\varepsilon} - vm_e\vec{v}$$

where v is the electron collision frequency. Re-writing the equation in terms of $\vec{r}$,

$$m_e \frac{d^2\vec{r}}{dt^2} = -e\vec{\varepsilon} - vm_e \frac{d\vec{r}}{dt}$$

We know that for sinusoidal fields, we can write equations in terms of phasors and replace d/dt with jω. Hence we can write the equation of motion on the electron in terms of phasors as,

$$-\omega^2 m_e r = -eE - j\omega v m_e r$$

Or

$$r = \frac{eE}{\omega^2 m_e - j\omega v m_e} = \frac{eE}{m_e \omega^2 \left(1 - j\dfrac{v}{\omega}\right)}$$

Substituting this into ($\vec{P} = -Ne\vec{r}$)

$$P = -\frac{Ne^2 E}{m_e \omega^2 \left(1 - j\dfrac{v}{\omega}\right)}$$

The electric flux density can then be found as,

$$D = \varepsilon_0 E + P = \varepsilon_0 E - \frac{Ne^2 E}{m_e \omega^2 \left(1 - j\dfrac{v}{\omega}\right)} \equiv \varepsilon_r \varepsilon_0 E.$$

The effective relative dielectric constant of the plasma is the

$$\varepsilon_r = 1 - \frac{Ne^2}{m_e \omega^2 \varepsilon_0 \left(1 - j\dfrac{v}{\omega}\right)}$$

An angular plasma frequency can be defined such,

$$\omega_p^2 = \frac{Ne^2}{m_e \varepsilon_0} \approx 3183N,$$

which is purely a function of electron density N. The,

$$\varepsilon_r = 1 - \frac{\omega_p^2}{\omega^2 \left(1 - jv/\omega\right)}$$

We see that in the presence of electron collisions, the dielectric constant can in general be complex. If we ignore collisions for the moment, then

$$\varepsilon_r = 1 - \frac{\omega_p^2}{\omega^2} \approx 1 - \frac{81N}{f^2}$$

From this result we can make several important observations. Since the propagation constant of a wave travelling in a plasma is $\sqrt{\varepsilon_r} k_0$,

- For frequencies $\omega > \omega p$, the effective dielectric constant is less than unity but the propagation constant is real. Hence, the wave will be refracted by the plasma according to the variation of ε_r with altitude.

- For frequencies $\omega < \omega p$, we get a negative value for the dielectric constant, which leads to an imaginary propagation constant. Hence, a plane wave in the medium will decay exponentially with distance. It is not absorbed (we have ignored losses/electron collisions here), but instead becomes evanescent, like a waveguide in cutoff. A wave incident on a medium with this propagation constant would be totally reflected.

- For frequencies $\omega \gg \omega p$, the effective dielectric constant is essentially 1. Practically this happens at VHF frequencies and above. The waves simply pass through the plasma without significant refraction, but there can be other effects, especially if the plasma is magnetized by the Earth's magnetic field and the medium becomes anisotropic. Waves at these frequencies undergo Faraday rotation by the ionosphere, whereby there polarization vector is rotated as the wave passes through the atmosphere.

From these observations we can observe two propagation mechanisms for a wave entering the ionosphere. If it is below the plasma frequency, it is simply reflected off the ionosphere. When it is above the plasma frequency, then refraction analogous to that which occurs with atmospheric refraction occurs, though it is much more pronounced because of the large changes in refractive index. In fact, provided the angle at which the wave enters the atmosphere is not too steep, the wave will be bent quite strongly as it enters ionosphere. If the angle of incidence reaches the critical angle for total internal reflection, the wave is reflected off the ionosphere and starts to return to earth, going through the reverse process of refraction as it does so.

It is possible, in the case of refraction, that the wave enters the ionosphere at too sharp an angle (close to broadside) such that the condition for total internal reflection to occur. Then the wave will pass through the ionosphere and never return. Hence, there exists a minimum set of conditions on the electron density, frequency, and angle of incidence, for the wave to be returned to earth.

If we subdivide the ionosphere into many tiny layers as we did for atmospheric refraction, we can say that,

$$n_0 \sin \theta_i = n_1 \sin \theta_1 = n_2 \sin \theta_2 = \cdots n_k \sin \theta_k \cdots$$

The condition for the wave to return to earth is to have total internal reflection, which begins when the refracted angle is $\theta = 90°$. If this happens at the kth layer,

$$n_0 \sin \theta_i = n_k \sin 90^0 = n_k$$

and since $n_o = 1$,

$$\sin^2 \theta_i = n_k^2 = \varepsilon_{r,k}.$$

From this, it follows that for a given angle of incidence θ_i and frequency f_{ob} (where ob stands for oblique incidence), the minimum electron density required to achieve total internal reflection is,

$$\varepsilon_{r,k} = \sin^2 \theta_i = 1 - \frac{81 N_{min}}{f_{ab}^2}$$

If the maximum electron density present is N_{max}, we can also view the condition for the wave to be returned to Earth as follows. In the most challenging refraction case, normal incidence ($\theta_i = 0°$, $\sin \theta_i = 0$), the only possible way for the wave to be totally internally reflected is if $\varepsilon_{r,k} = 0$. This requires the frequency to be less than the critical frequency f_c, given by,

$$\frac{81 N_{min}}{f_c^2} = 1 \Rightarrow f_c = 9 \sqrt{N_{max}}$$

If the electron density present is N_{max}, ($\varepsilon_{r,k} = \sin^2 \theta_i = 1 - \frac{81 N_{min}}{f_{ab}^2}$) can be rewritten in terms of the critical frequency as,

$$\sin \theta_i = 1 - \cos^2 \theta_i = 1 - \frac{81 N_{min}}{f_{ab}^2}$$

$$f_{ab} = 9\sqrt{N_{max}} \sec \theta_i = f_c \sec \theta_i$$

This value of f_{ob} is called the maximum usable frequency, and is less than 40 MHz, and can be as low as 25-30 MHz in period of low solar activity. Equation $f_{ab} = 9\sqrt{N_{max}} \sec \theta_i = f_c \sec \theta_i$ is called the Secant Law.

Figure below shows the curved path of a refracted ray associated with frequency f_{ob}. The curve path reaches an altitude of h_1 before being returned to the Earth. If the incident and returned rays are extrapolated to a vertex, they meet at a height h' which is called the virtual reflection height of the ionospheric layer. The virtual height depends on conditions, the time of day, and the layer being considered, as shown in table.

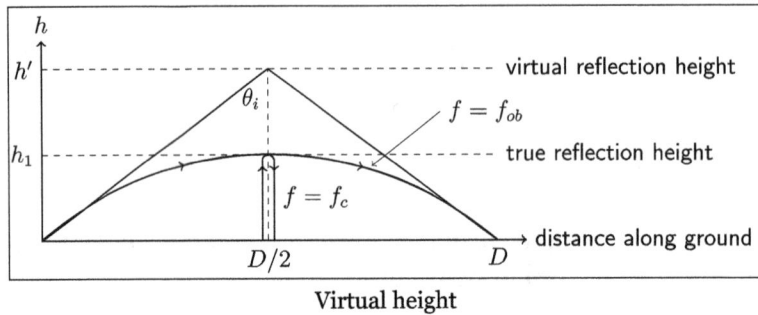

Virtual height

Table: Virtual heights of various ionospheric layers

Layer	Daytime virtual refl. height	Nighttime virtual ref. height
F$_2$	250-400 km	—
F$_1$	200-250 km	—
F	—	300 km
E	110KM	110 km

From the geometry of figure, the distance traversed by the curved path can be determined from,

$$\sec \theta_i = \sqrt{\left(\frac{D}{2h'}\right)^2 + 1}$$

The maximum skip distance $D = d_{max}$ can be achieved by aiming the radiation from the antenna so that the radiation leave the antenna parallel to the Earth's surface. This situation is shown in figure. Factoring in atmospheric refraction, the maximum skip distance is given by,

$$d_{max} = 2\sqrt{2K R_e h'}$$

which shows that very large propagation distances are possible, especially when the upper ionospheric layers are used. For example, if the F layer is used, using an effective Earth radius of KR_e = 8497 km gives a skip distance of 4516 km. Multiple skips are possible by using reflections of the earth to establish a multi-reflection process. This equation explains why signals can propagate so much farther at night when the D layer disappears, since it has the lowest virtual height.

In general, the broadcaster can limit the range of the transmitter by controlling the power and by manipulating the angle of radiation. Pointing the antenna skyward reduces the angle of incidence, resulting in a shorter skip distance.

References

- Kimmel WD, Gerke D (2018). Electromagnetic Compatibility in Medical Equipment: A Guide for Designers and Installers. Routledge. p. 6.67. ISBN 9781351453370

- Wave-propagation-definition-equation-and-types: elprocus.com, Retrieved 31 July, 2019

- "Agents Classified by the IARC Monographs, Volumes 1–123". Monographs.iarc.fr. IARC. 9 Nov 2018. Retrieved 9 Jan 2019

- Seybold JS (2005). "1.2 Modes of Propagation". Introduction to RF Propagation. John Wiley and Sons. Pp. 3–10. ISBN 0471743682

- "Why AM Stations Must Reduce Power, Change Operations, or Cease Broadcasting at Night". Federal Communications Commission. 2015-12-11. Retrieved 2017-02-11

- What-is-space-wave-propagation, principles-of-communication, physics: thebigger.com, Retrieved 31 July, 2019

Permissions

Index